NATURE at NIGHT

NATURE
at NIGHT

Discover the Hidden World That Comes Alive after Dark

Charles Hood

Timber Press
Portland, Oregon

Timber Press
Workman Publishing
Hachette Book Group, Inc.
1290 Avenue of the Americas
New York, New York 10104
timberpress.com

Timber Press is an imprint of Workman Publishing, a division of Hachette Book Group, Inc. The Timber Press name and logo are registered trademarks of Hachette Book Group, Inc.

Printed in China on responsibly sourced paper
Text and cover design by Leigh Thomas

The publisher is not responsible for websites (or their content) that are not owned by the publisher.

The Hachette Speakers Bureau provides a wide range of authors for speaking events. To find out more, go to hachettespeakersbureau.com or email hachettespeakers@hbgusa.com.

ISBN 978-1-64326-313-7

A catalog record for this book is available from the Library of Congress.

Contents ⦾⦿⦿●⦾●⦾⦾

Introduction

Welcome to night—the least studied, least photographed, least appreciated segment of our planet's nonstop, twenty-four-hour go-go show. In this book we'll look at how nocturnal nature works around the world, from the vertical migration of squid in the open ocean to the hunting strategies of a bat that targets millipedes and scorpions. We will dance with fireflies and learn how the jungle experts photograph ocelots. (Hint: try spritzing the trail camera with men's cologne.) One intrepid animal we will consider is the globe skimmer, which is also called the wandering glider. This is a dragonfly whose transoceanic migration route follows the monsoons from India to Africa and back again, with a stopover in the Maldives. No single individual dragonfly makes the full journey; instead, successive generations use trade winds and rain-fed pools to propel the population on a global journey.

The globe skimmer has a multigenerational migration route that covers over eleven thousand miles.

The skimmer's story reminds us that the world is richer, more diverse, more interesting, and stranger than many of us might guess. Most of us have a basic idea of "daytime" nature—even if you've never been on a safari, you already know how to tell a zebra from a gazelle or a cheetah from a jackal. Nocturnal nature is much less commonly known, and some of us are even a bit hesitant about hiking after dark. After all, if we go out at night, won't we step on a snake or get dragged off by wolves? (Quick answer is "no" to both.)

Our hesitation may be because night is still associated with danger and immorality. Describing London before streetlights, night scholar Sukhdev Sandhu reminds us that "night was seen as lawless, foreign territory teeming with rogues and banditos who took advantage of what Shakespeare called its 'vast, sin-concealing chaos' to revel in an orgy of depravity and moral pestilence." The historical truth was neither so grim nor so morally questionable, and in terms of numbers, there was just as much infidelity happening during the daytime as there was after dark. London's streets began to be lit with gas lamps starting in the early nineteenth century; this also was when more effective

In Hindu mythology, the Moon is a male figure, Chandra, shown here in a coach drawn by a blackbuck antelope.

lighthouse lenses were developed and rocky coasts could be marked at night to make shipwrecks less inevitable.

If we fear the night today, it may be because we no longer remember our once-intimate relationship to it. The endless generations of humans who came before us moved through darkness easily, and in doing so, they saw more, smelled more, and heard more than many of us do today. They knew how to *listen*, and because of that, they knew that the night could be beautiful—beautiful visually but also beautiful aurally. According to English professor Michael Warren, "In the dark, sylvan villages of medieval England, people named places after the birds that filled the night with music." Among other examples, he mentions Snitterfield (named for a nocturnal shorebird, the snipe) and the tiny village of Ulcombe, "nestled in a gentle fold of the North Downs." The place name means "the owl's valley." Which kind of owl? Maybe the tawny owl, maybe a long-eared owl, maybe a barn owl—lots of possibilities, then and now.

Anglo-Saxon poetry remembers the wild sounds of gannets and curlews; Chaucer cites the grunts, pumps, and booms of nighttime bitterns. The Japanese haiku master Kobayashi Issa has a very moving piece about an itinerant monk trying to cross a dry creek bed after dark. The monk has no lantern and a storm is coming. As we all know, flash-flood danger means to stay out of arroyos during rainstorms. Yet the unnamed walker has lost the trail and is now trapped in the creek, trying to find his way using just the occasional flashes of lightning. Does he make it? We assume so, since the poem seems to describe a personal experience. The lost walker saw enough each time to move forward a few feet to get out of the riverbed. Ultimately, he must have come

to a village. The poem is so compressed that it comes and goes like a lightning flash itself.

This is an interesting text but not singular. In fact, night has a literary or artistic presence in almost every culture we can name, and it often can be a very potent force. The Egyptian goddess of the night, Nephthys, is often shown with falcon wings, while in Maya cosmology, night is personified as a jaguar. Hindu's moon lord is Chandra; his chariot is sometimes drawn by an antelope and other times by seven white geese. The Catholic church venerates Peter of Alcántara, the patron saint of night watchmen and the figure most closely associated with the nocturnal adoration of the Blessed Sacrament. To spend more time in prayer, the saint slept very little; among his other miraculous virtues, he could reportedly levitate.

Night generates its own vocabulary. Victorian poetry borrowed "gloaming" from Scots English as another way of describing "twilight," and the word became a common literary expression in the nineteenth century. Herman Melville in the 1860s, for example, said that fear comes into our minds "like the thief in the gloaming." ("The Gloaming" is also the title of a song by the contemporary band Radiohead.) On average, that word excepted, English lacks good options for describing early evening. In French there is a lovely term, *l'heure bleue*, which means "blue hour," that late twilight hour when the sky is filled with a serene, blue light. That phrase is just right—it even sounds serene when you say it quietly to yourself.

In all but the most recent times, humans intimately knew the stars and planets, and they were such striking objects that people often attributed agency to them. In medieval cosmology, to be "under the sign of Saturn" meant a person was melancholy, lacking in purpose. Susan Sontag plays on this with the title of her book, *Under the Sign of Saturn*, and critic Walter Benjamin alludes to it as well. Where these previous authors alluded to moods and hindrance, other people now celebrate the outer

planets as large, successful protostars, filled with possibility and wonder. Jupiter has green lightning; Neptune takes 165 years to make a full orbit around the sun; Saturn's moon Enceladus has an icy outer crust protecting a liquid ocean within. Recent discoveries confirm that the slushy, salty, unfrozen churning ocean on Enceladus contains the same chemical markers that make life possible on Earth. If we want a chance of finding other life forms in the solar system, Enceladus is a good place to start looking. Notes on this occur on page 268.

And then there is the moon itself, our four-billion-year-old companion. A large planet like Jupiter has between eighty and ninety moons, depending on where one draws the boundary line between "rather small moon" and "large, orbiting boulder." We now know that even far-off Pluto has five moons. On Earth we have just the one, but it has fired imagination as far back as we can trace language. The day of the week that follows Sunday—the much-maligned Monday—takes its name from the Anglo-Saxon words for "day of the moon." ("Tuesday" comes from a word for the planet Mars.) An appreciation of the moon unifies cultures across the globe. And in fact, the base words that became our modern words "lunar" and "moon" date back to the earliest glimmers of proto-Indo-European, and as such, they are among the earliest words we can document in any language.

From ancient times until now, the full moon has never left our imagination. Lay it on us, Dean Martin: "When the moon hits your eye like a big-a pizza pie / that's *amore*." Don't you wish you could have a dollar for every time moon or moonlight turned up in a poem or song? Maybe songwriters like the moon because it provides such easy rhymes: swoon, croon, rune, a tune in June, *Brigadoon*, Vidal Sassoon. Other parts of the night sky have inspired music as well. In "Rocky Mountain High," when John Denver sings "I've seen it raining fire in the sky," he is sharing his memory of the Perseid meteor shower.

Why Do We Have Night in the First Place?

The Earth rotates on its axis—one half always faces the sun, and one half faces away—and at the same time, it leans over in a strong tilt, so seasonally, on the year-long circuit around the sun, one part of the planet has very long days. When that is happening, the other part has long nights. (The equator has more or less the same length of day and night all the time.)

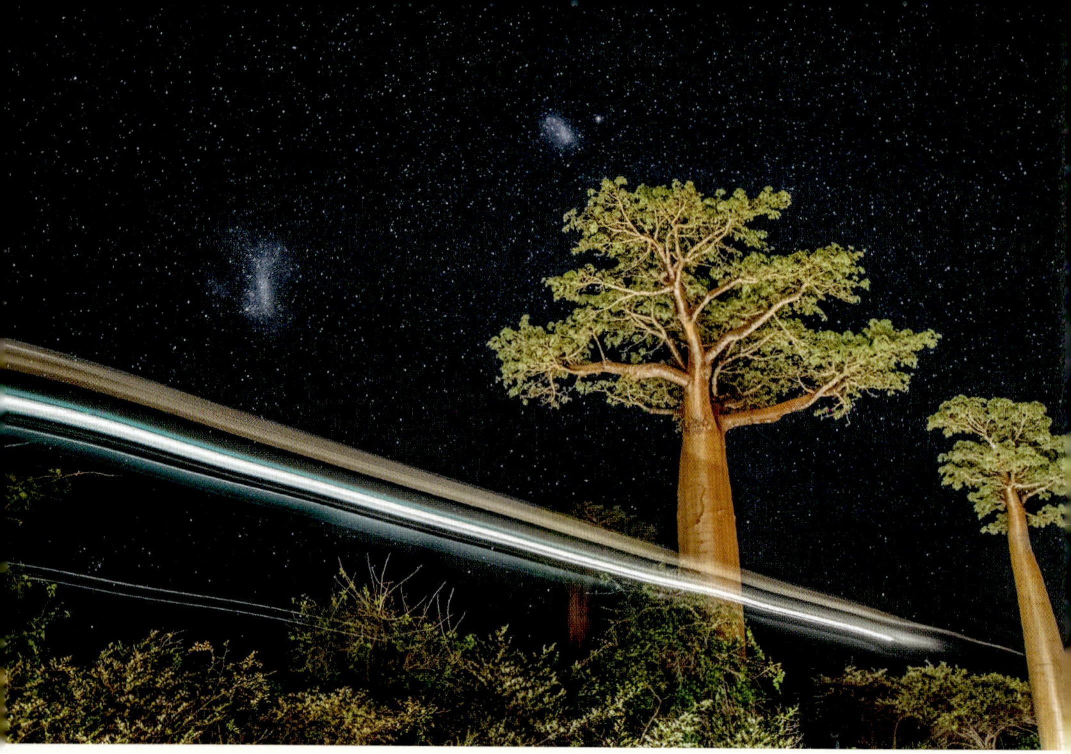

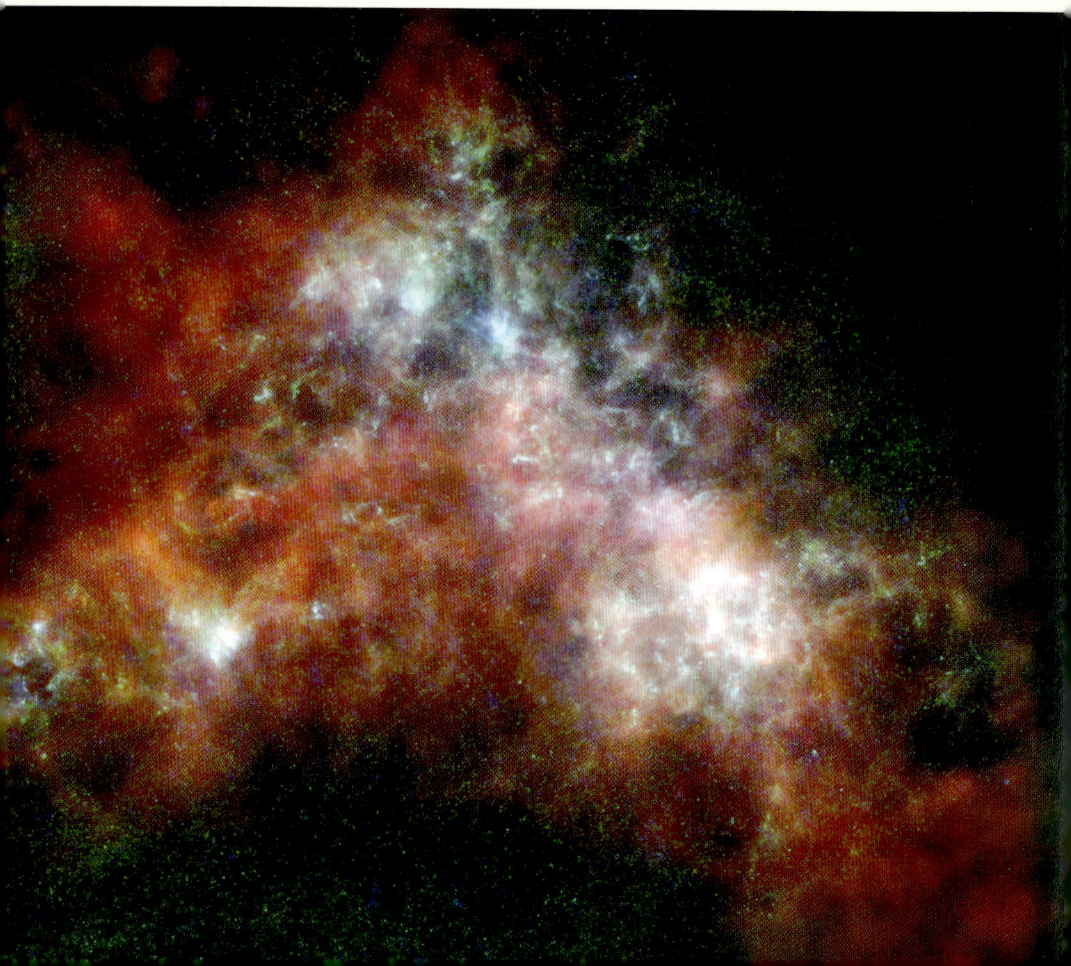

◄ Baobab trees in Madagascar are lit from above by the Magellanic Clouds and from below by passing trucks.

All of this feels normal to us, and most religious traditions have mechanisms to explain who or what created "day" and "night." If we lived someplace else, the pattern would be different. On the surface of Mercury, the sun would look three times larger than it does from Earth, and the sun's light would be seven times brighter. Yet Mercury has an eccentric orbit and a very slow rotation, which creates different diurnal cycles than what we're used to. According to NASA, on some parts of Mercury, "The morning sun appears to rise briefly, set, and rise again ... The same thing happens in reverse at sunset for other parts of the surface."

Our own nighttime experiences vary as well. What you see in terms of the patterns of the stars depends on where you are standing as you look up. In the southern hemisphere, not only are the seasons reversed from the northern pattern (making Christmas a summer holiday), but sky features differ, too. Things you can only see at night from the southern part of the planet include the Magellanic Clouds, two irregular galaxies that share a gaseous envelope. These "clouds" are dwarf galaxies formed by a few hundred of the many thousands of stars that orbit the Milky Way. These "add-on" galaxies are influenced by (and linked to) our galaxy's gravity in the same way that the solar system's planets orbit the sun. Seen on a clear night, Magellanic Clouds look like somebody has dipped a paint brush into the Milky Way and splotched some of the extra color onto a blank part of the sky.

◄ The Small Magellanic Cloud is a dwarf galaxy orbiting the Milky Way. This view combines space-based and land-based observations into a single image.

You can't see these features from Los Angeles or Paris, and not just because of light pollution. Each part of Earth looks out into a different piece of the universe, and so, like the more famous Southern Cross, Magellanic Clouds are special features of the southern hemisphere—they are visual experiences denied to "northerners."

Since specific views are dependent on location, our earliest ancestors looked up in fear and awe at a midnight sky that was different from the one you grew up with (unless, of course, you grew up in East Africa). For more on our primate cousins, see the night monkeys chapter.

Why Are Some Animals Out Only at Night?

Everybody knows that bats come out at night, but why? Why are there not flocks of daytime bats the same way there are gaggles of geese or murmurations of starlings? One answer is that there are: If you ever get the chance to stand near (but never directly *under*) a roost tree of the kind of fruit-eating bats called "flying foxes," during the day they do jostle and shift and even lift up into the sky to change trees or survey intruders. (They also rain down a constant stream of poo.) And even strictly "nocturnal" species of bats may come out early in the evening when it is still fairly light. They might hunt for a few hours around sunset, rest in the very middle-most part of the night, and then hunt again before dawn. The midnight resting sites are called night roosts and often differ from the daytime roosts. You can sometimes find bat night roosts during the day by looking for bat guano. This residue of droppings accumulates at sites that are used often, and if you find guano, then later in the night you might find a midnight roost of bats.

Do bats come out during a total solar eclipse? Usually not; it's too brief a period of time and their body clocks know it's not yet time to get up. A lunar eclipse is different. A darker night may be better for bats, whose echolocation skills give them an advantage—and even a brief amount of extra darkness is going to be something they can exploit. This fact still does not explain the question, "Why use night at all?" Most birds,

A pygmy kingfisher sleeping on a vine in Borneo looks like a fuzzy neon ball. The beak and head are tucked under the wing.

for example, roost at night, like this sleeping kingfisher in Borneo.

But exactly because most birds are asleep at night—the lack of light making it too hard for visual gleaners to forage for seeds, and (in some groups) the cool temperatures making insects less active—this means that for other groups, like mice, conditions are more favorable. On average, if you're a mouse, it's easier to stay alive at night than it is during the day. While mice may have owls and foxes to deal with at night, they do not have hawks and angry farmers. And being warm-blooded, mice can be active in early spring and late fall when temperatures are too low for most snakes to be around to hunt them, since snakes need the sun to come up and warm them before they can hunt. If you can see well in the dark and navigate the world by smell, the nocturnal lifestyle makes a lot of sense.

A nocturnal lifestyle makes sense for insects, too, although for different reasons. Things are around at night that eat insects (owls, bats, toads), but it's still easier to make a living in the darkness than it is in harsh noon light. It's all a matter

A turquoise jay in Ecuador proves that the early bird catches the moth.

of being able to hide from predators . . . and being able to find things to eat. Sometimes the timing doesn't work out quite right. If a night-flying moth (for example) misses dawn's cues and stays out into the full morning, that moth may end up becoming a blue jay's breakfast. In the photograph here, a jay in Ecuador has arrived at dawn to investigate a farmer's "left-on-all-night" porchlight, to see if any insects have stayed out past their bedtimes. The answer this time was "yes," making for an unlucky insect but a lucky bird—and the circle of life continues.

Some "day" animals have large sizes and large appetites, and so they will forage all night. If there are no poachers or other outside influences, elephants, for example, can feed both day and night, as will bears and whales. The story, as always, is complicated. Blue whales off the coast of California might rest at night (and thus become vulnerable to ship strikes) if their main prey, the swarms of tiny krill, dissipates. Yet in the same location, humpback whales will feed at night if they find good concentrations of fish. If this happens, the humpbacks will have time (and energy) during the day to carry out the social interactions that delight people on whale watching trips—all of that body-lunging and flipper-waving that makes humpbacks easily visible and popular.

A typical compromise between night and day is enacted by the western spotted orbweaver, a kind of spider found from the middle of North America south into central South America and the Galápagos Islands. The species name—*oaxacensis*—shows us that when it was described, it was thought to be only from Oaxaca, in south-central Mexico, which is about the midpoint of its range. These spiders use their webs to snag flying insects like beetles, moths, flies, and bark lice; to be out in the daytime not only means they might get discovered by a wren or jay, but also means they risk getting cooked in the sun. Spiders can overheat just as humans do, so during the day they often retreat to the shaded edges of their webs, sometimes hiding inside curled

Western spotted orbweavers have attractive body designs. Each individual spider's pattern is different.

leaves. Prey may become trapped in a web during the daytime, but usually the spider waits to eat it. Spiders need the night; there is more prey overall then, and it is also the safest time for them to be out in the open, ready to repair their webs and to eat whatever has been trapped in them.

Why Study Nature at Night?

Just because bugs and bats are out at night, that doesn't mean we need to do anything about it. Some folks might say, "Good: Let all those creepy things stay on their side of the clock and I will stay on mine." Part of that night-fear spoken of earlier gets transferred into a hostility toward the wildlife that inhabits night. This is neither fair nor logical, but it's definitely a common response. In an informal survey, I've found that most people like butterflies (bright, diurnal) but have much less generous feelings toward butterflies' cousins, the smaller, darker, more "fearsome" moths. In practical reality, there's hardly any difference. To quote insect researcher Eric Metzler, "Butterflies are just moths in fancy dress."

Yet that's not how they are perceived. In the Godzilla franchise, the other evil animal is Mothra, and Mothra is based loosely on a silk moth. The other monster is not called "Monarch-Butterfly-Ra" or "She-Fang, the Giant Swallowtail." Put "moth" in the phrase and you're already halfway to "evil" on the white hat / dark hat sliding scale. There is even a moth whose common name is the "death's-head hawk moth." That sounds so grim, doesn't it? There's a goth band called "Coffin Moth"—no day-flying butterfly ever gets a name with connotations like that.

If we set aside our distrust and let nature at night come to us on its own terms, we are going to be given great gifts. If you have a good sense of smell, there is no perfume shop in the world as rich and powerful as being in the rainforest at night. Smells are

The coronated treefrog lives in rainforests in Costa Rica.

hard to describe, especially aromas that we have not encountered before, but people talk about the forest as "rich" or "earthy" or "alive." *Compost* and *humus* are also words associated with the smell. It's more than that, and sometimes the range of smell is almost overwhelming if there is rotting fruit on the ground or plants like skunk cabbages around. (For a plant that smells like rotting meat, Borneo has the famous "corpse flower," which is also cultivated in some botanical gardens in Europe and North America.)

To give you a sense of the possibility of all that you can smell, remember that vanilla was originally native to the New World tropics; cinnamon to India; cloves and nutmeg to Indonesia. These are all wet-forest plants. As a side note—and we will get to this later—vanilla plants are bat-pollinated, as are bananas, avocados, papayas, and the agaves used for tequila.

Or maybe you just want to see cool stuff, skipping perfumery and not caring about the buzzy buzz of the cicadas. Here's just one thing picked out of ten thousand: the coronated treefrog.

This is a Central American rainforest species, so it has a range from Mexico to Panama. It is nocturnal, it lives in the water of bromeliads, and the male's *bop bop bop* call can be heard from three hundred feet away. Given that the frog is just three inches long, the range of that call, if scaled up and applied to the average person, would mean that if you wanted to belt out your favorite sing-in-the-shower song, your voice could be heard a mile and a half away.

The only way to see this frog is to go out at night. (Well, you might see one in a zoo or at a pet show, but in this book, we give priority to "in the wild" experiences.) The thing is, you do not need to go on an expedition to another country to have great encounters. Once we start to look, we can find "nature at night" everywhere. We have a photo below of a lacewing that has landed on a girl's finger. This did not take place inside a national park or next to an elite nature reserve. It was at a rest area on the highway—just a case of leaving the car window open and letting nature come to us.

If you have flipped through this book much, you've seen that we're going to talk about glow-in-the-dark mushrooms and do a detailed account of a night walk in Panama. This will help us go much deeper into science topics than a regular hiking guide can.

This delicate lacewing landed on someone's finger at a freeway rest area.

Compare nature to food: As necessary as the plain foods might sometimes be—white rice, saltine crackers, store-bought tortillas—most of us are happier if we can also have color and zest. Bring on the homemade blackberry cobbler; please pass the mango salsa. Being more open to the sounds, smells, sights of nocturnal nature simply means getting extra helpings of nature overall.

Equipment for Enjoying Nocturnal Wildlife

To start with the basics, get a good flashlight, and then get a second one as a spare. Many hikers like to use a headlamp as one of their lights. This helps keep one's hands free, and no matter what style of light, choosing one with a red setting can be good. It helps preserve your night vision while you do simple chores, it bothers skittish wildlife less than a brighter light does, and it attracts fewer bugs to hover annoyingly around your eyes. In areas with a British or Commonwealth influence (including safari lodges in Africa), a small, handheld light is sometimes called a torch.

Kids love running around waving flashlights in every direction—adults do too, if they are willing to admit it—but try to convince everybody to aim down, not up. It takes several minutes for our eyes to adapt to a dark night, usually in several stages. There's a "phase one," and that happens right away, in just a minute or two. But our final night vision takes up to thirty minutes to arrive. Many of us never experience this, since even a movie theater has a bright screen and well-marked exits. And because most cities are so over-lit, it's hard to appreciate just how good our eyes really are. Yet give yourself a chance, and your eyes *will* adapt: All humans have the potential for excellent night vision—it's how we evolved. You can even hike using only starlight, though we still want you to have your flash-light with you anyway. (See the note about common sense, page 33.)

The problem with this superpower is how easily it gets switched off. Your night vision is quickly blown out when somebody aims a flashlight at your face, and the same thing happens when a late-arriving com-panion's high-beams scald the parking lot. Your recalibration has to start all over again. Jane Slade, an expert in outdoor lighting, defends the role that a dark sky can play. She sees a night walk (or even a "night sit") as a time for rest and recovery. If there are too many lights on at once, she says, "it's not just a loss of star-gazing; it's a loss of self-gazing." From her perspective, "darkness is a birthright for the natural world." So that means whenever you're outside, dim your vehicle lights (including the interior dome light in the car), keep your voices down, and, for kids of all ages, please practice some restraint when it comes to waving a flashlight like a Sith lightsaber.

Specialty tools for exploring the night range in price and may stay on your wish list for a while. The cheapest is a basic UV flashlight, useful for summer scorpion hunts. Bat detectors plug into your phone or are stand-alone units about the size of an old-fashioned transistor radio; these are expensive but make bat echolocation calls accessible both visually and aurally. In Europe and North America, some detectors also offer ID sug-gestions (with varying degrees of reliability). Truly hardcore night-ologists covet thermal scopes, which read differences in ambient temperature. Hot things (such as animals) glow white on these scopes compared to the cooler foliage around them. Most models are monocular and are shaped like small camcorders from the VHS days. Using a heat scope, you can see a tropical por-cupine in a tangle of jungle vines, a Eurasian woodcock foraging

The sign on this observato-ry's gate tries to protect the night vision of guests and workers alike.

Bats, like birds, have unique "songs." This readout shows the echolocation signal of a Yuma myotis bat.

on the forest floor, or even a migrating whale's warm spout if you're doing a nighttime cetacean survey. (Whales are so well-insulated by their blubber that the animals themselves don't usually show up, but their spouts are visible because their breath is almost always warmer than the surrounding air.)

Sometimes you end up using all of these tools in one night, which makes for an interesting experience but a heavy pack. That's especially true if you add a camera or two, a long lens, and some flash units.

If you want to know what's using a nearby trail or passing through your backyard at night, another option is a trail camera. These are weatherproof boxes with infrared triggers that can produce a grainy but decipherable wildlife image. See, for example, the photo on the next page of a Virginia opossum caught by a trail camera in the Santa Monica Mountains. Trail cameras can be set up to record video, too, and image quality improves each year. Trail cameras in southeastern Arizona have documented

Researchers near the Grand Canyon are catching bats in nets (far left), measuring them on the worktable (middle), and getting supplies from their trucks (far right). The Milky Way stretches over everything.

jaguars, peccaries, bears, nudists, and drug couriers backpacking with bales of marijuana. For higher-quality images, my partner and I use secondhand cameras and dedicated flashes. To do this, my photo buddy, José Gabriel Martínez-Fonseca, MacGyvers old plastic boxes and extra battery packs to create long-duration survey packages. The below image is one of his shots from his native country of Nicaragua, showing a margay—a small jungle cat—triggering a beam that takes its picture. We will see more of José's work in the bats and small cats sections of the book. (He also took the Milky Way shot on the previous page.)

The Virginia opossum was introduced to California in 1910. It is captured here by a trail camera in the hills near Los Angeles.

Night doesn't end when the sun rises. Even if a person has spent the night resting in a large, comfy bed, our nocturnal study can still happen anyway. An early morning walk on local trails allows you to survey scat, footprints, and sometimes even the remains of a kill, and that can reveal the drama that happened the previous night. In a snowy Yellowstone winter, a wolverine's tracks in the snow mark where it left its den and headed out to forage. These hardy animals do not hibernate, and since they have broad paws, dense fur, and sharp claws that are good for

Old camera bodies never die, they just get put in plastic boxes and strapped to jungle trees. This small cat, the margay, was photographed in Nicaragua using a modified DSLR camera.

Tracks in Yellowstone National Park reveal a wolverine's route. You don't always need to see an animal directly to know that it passed nearby.

climbing trees and icy slopes, wolverines easily traverse winter landscapes. They sometimes use cached food from the earlier months, and their strong jaws can crack open bones, should they find a long-dead moose. *Pugnacious* is one word used often to describe them; most of the time, a wolverine is even willing to stand up to a bear.

There are field guides and websites that can help you identify the scat (animal poo) that marks a morning trail. Some animals (rabbits, deer) just go as the need arises, while other animals (raccoons, wolves) scent-mark intentionally, using latrine sites as boundary markers. These can be very helpful to passing naturalists, since they can be just as reliable as a trail camera in helping to document which animals are using the trail. On a camera, images can be blurred or just show a tail tip, but a pile of animal dung never lies.

"What a Beautiful Jay, Jay"

Time for a brief note on the arcane science of naming animals. The beautiful jay (*Cyanolyca pulchra*) lives in mossy, mid-altitude forests in Colombia and Ecuador. Some ornithology handbooks would write its common name as "Beautiful Jay" (with capital letters), to distinguish between it and all the other attractive jays in the world. By that convention, there are many beautiful jays, but only one Beautiful Jay. That makes sense, but how do we transfer that model to other groups? Not to be slighted, some mammal watchers use capital letters too; they would say the small cat on the previous page is not a margay, but a Margay. (The word *margay* comes to English from Spanish, which borrowed the term from Portuguese, which appropriated it from a now-extinct tribal language, Tupi. Our mouths breathe out little smoke puffs of history every time we speak a sentence.)

The problem is, such a system of capitalization becomes cumbersome once we include other life forms. The editors and I want to spare you the experience of trying to make your way through a passage like this: "The Lodgepole Chipmunk in this photograph is scolding Human Hikers from the bough of a Ponderosa Pine, next to a meadow filled with Mountain Coyote Mint and Woolly Mule's Ears. A small flight of Mountain Emerald Dragonflies hovers nearby." It all gets to be too much very quickly.

"I'm just a chipmunk. Don't drag me into your language wars!"

Proper names should be capitalized—all style guides agree on that—so on page 26, that animal is a Virginia opossum. Other than those instances, in this book, all animals and plants will be written the usual way—coyote, mermaid, peacock, oak tree.

Where To Go

This is easy: go anyplace you can, any day of the week. Is a rich uncle going to send you on an African safari? Great, pack up and go. Night "targets" in Kenya include servals (small wildcats), bush babies (tiny nocturnal monkeys; page 224), and a solitary, night-loving plover, the violet-tipped courser. It's also called the bronze-winged courser, but I prefer the first name. Everybody knows Africa is good for nature study, but just how good is it? Well, the world record for the most mammals seen in one day was set in East Africa in 2021—sixty-six species seen in twenty-four hours. For birds, experts can tally up seven hundred species in a few intense weeks. So yes, from bugs to baboons, Africa is an easy choice.

On the other hand, you don't need to go anyplace special. In the aerial photograph above, we're in a helicopter looking over

Downtown Los
Angeles has
more nature
than most
people expect.

downtown Los Angeles. This is about as "urban" as urban can get, and yet—

(we pause now to put our prejudices back on the shelf)

—rather than being a concrete wasteland, within five miles of the center of the location, there are (or very recently have been) bobcats, coyotes, raccoons, skunks, foxes, owls, the largest and smallest bat species in North America, and at least one puma, the famous P-22 of Griffith Park. As we go to press, its home range remains empty. P-22 died recently due to a combination of old age and the wounds from being hit by a car. This night view of Los Angeles is all animal territory. The last wild grizzly bear in Southern California was seen in a grape orchard just north of this photograph, and black bears turn up in nearby suburbs.

So where can we go to start nature study? Our own yards and villages and cities and home blocks, that's one good answer.

From Singapore to Lisbon, from Anchorage to Patagonia, there's not a city anywhere that doesn't have some combination of ducks and hawks, lizards or geckos or frogs or salamanders, squirrels or rabbits or mice or strange little blurs nobody has bothered to identify yet. A short drive from downtown Anchorage and you can look for beluga whales in Cook Inlet; I've seen red foxes in London and Helsinki and downtown Fairbanks; Paris has feral parrots, even in the middle of winter. Mumbai has fifty leopards in the main city and another few hundred in the surrounding area. We know that India is now the world's most populous country, which puts pressure on all of its wildlife. And yet I once had a "five canid" day in India, meaning I saw five dog species in the same day. The five were wolf, dhole (also called Indian wild dog), golden jackal, Bengal fox, and escaped village dogs that had formed their own wild pack. Do feral animals still count as wild and interesting? They do to me—I love seeing wild mustangs and flocks of urban parrots, and I have argued elsewhere that I wish there were more escapee monkeys in the world. And so, five species of dog in one day may not seem like a lot, but you can't see that many even in Yellowstone or the Amazon. Usually, two or three is the max species count per day. India has tremendous cultural riches, yet the wildlife diversity is equally impressive.

Thinking about the leopards of Mumbai and the coyotes of Chicago allows us to come back to one of several ideas this book explores. When it comes to bad news about the environment, it's hard to know what to list first. This will be the first of only two instances when the book will use the expression "climate change," but the facts are the facts, and as the century progresses, seas will rise, glaciers will thaw, and if we were ever to find evidence of life on Mars, no sooner will that be announced

Coyotes have survived all attempts to eradicate them and can be found from Winnipeg to Miami and across all of the southwestern United States.

than there will be a follow-up press release: "Life found on Mars . . . and it is already contaminated with microplastics." And yet despite all the bad news, there is reason for hope. Good people make good progress every day, on everything from regenerating coral reefs to reducing household food waste. Once-lost species are later rediscovered, and range maps get extended every week. There is much we don't know, and there are many curious minds working hard to figure things out.

Rather than bringing up all the bad things going on right now, this book celebrates the good things: the interesting places, the exciting discoveries, the new connections that scientists (and artists and poets) make every day. There was never a time on the planet when people were not being stupid or mean, and in that sense, our age is no different from any other age. And yet separate from that, this is such a golden time for nature study. We have so many tools—from cameras to satellite maps to top-notch field guides—that it's easier than ever to see things well and to document what you've seen in order to share with others. If you're not using iNaturalist (self-described as "an online social network of people sharing biodiversity information") yet to post your sightings, we hope after your first few trips you will start. We will talk more about the pleasures of community science later in the book.

The rhinoceros chameleon has eyes that swivel independently.

We see on the previous page a great picture of a rhinoceros chameleon. Speaking as somebody who started out using slide film, first off, can I just praise the heck out of the photographer here? This shot, taken in the middle of the night, is *so clear*. My co-conspirator, José Gabriel, has the focus and the flash power dialed in just right. More important, this shot was taken behind our cabin. José and I were in Madagascar, but we had not made a special trek to the deepest part of the swamp to find this amazing creature. We did not have leeches hanging off the tips of our noses; there were not galloping herds of vipers about to latch onto our bums. If we had wanted to, we could have walked from here five minutes back to the open-air kitchen and gotten two cold beers.

The take-home message? Nature is cool; nature is all around us; nature says that you just need to give yourself permission to go and look, and if you do, the world will offer up a nonstop cascade of surprise and joy. This chameleon (like others of this genus) can swivel its eyes in their little gun turrets up, down, to the side. Look at that snout! It's as if it has a fake chainsaw sticking out of its face. I love the graceful curl of the tail and the way the fingers cling to the branch. And boy, does this fellow look grumpy. He does *not* want his picture taken. That's okay, pal, a few quick shots and we will be on our way.

This species is endemic to Madagascar and can grow to be nearly a foot long. It zaps its insect prey with a quick flash of its sticky tongue.

Let's Be Safe Out There

Four words: *Have fun, be safe*. Never go hiking at night without a reliable light and at least one backup. (Two or three extra lights would be even better.) Tell people where you're going and when you'll be back. Take a buddy. Watch where you're stepping. If a place feels sketchy, trust your gut and turn back early.

Use waypoints to mark your route so you can find your way back, and look behind you, just to see who (or what) might be following. This is all normal advice for being outdoors anywhere. In any kind of hiking—daytime, nighttime, or in between—common sense goes a long way.

One of the author's field companions, Paul Carter, has seen even more animals than the author has, so we will turn to him for advice about staying safe. Paul lives half the year in Thailand and also spends time each year in California. Those are just his home bases; from those sites, he travels all over the world. He is especially interested in nocturnal snakes and mammals, and in the past thirty-five years, he estimates he has completed a thousand nighttime excursions.

How many times has he died doing all these nighttime rambles? So far none, touch wood. He does take precautions, though.

"After I once planted a foot within inches of a five-foot pit viper in Costa Rica, I now wear snake gaiters and leather boots, as opposed to shoes with thin mesh and suede uppers. And my GPS unit is generally on, with the track log activated. It's easy to become disoriented in forests and hills when tracking animals away from the path. I also use a GPS to mark waypoints for sightings that are worth reporting in iNaturalist."

He adds, "Apart from the obvious items such as binoculars and camera gear, I always have a thermal scope with me. For lighting, I take a headlamp as well as a handheld spotlight. I prefer my lights to have a tightly focused beam so that I don't light up the entire forest and/or any nearby housing. My headlight gives me enough flood beam when that is needed."

Spare batteries? Check. He also has first-aid supplies and plenty of water. You need to be sensible and take the usual care. He suggests people put things in perspective. "Depending on where you are, I would say that there is probably an increase in risk when going out at night compared to daytime hikes. But the level of danger is less than the common perception."

How do we explore wildlife at night and yet stay safe? Paul reminded me, "The best way to reduce apprehension is to check the area in daytime and then to go with a companion after dark. I do a lot of night hikes in Asia, and some areas are certainly no-go zones because of carnivores or elephants. So besides common sense, it is as important to be as aware of your surroundings in the night as it is in the day. And walks at night become more manageable once your comfort levels and field skills increase with experience. That's like everything, really."

As a pale-skinned survivor of melanoma, I joke that I like to study nature at night since you can't get a sunburn. I also like the way a flashlight helps center my attention: It turns us all into poets, reaching out with our imaginary spoons to sample the world one serving at a time. This book hopes to show what happens in the shadows during the planet's hidden hours. There is a lot going on, though much of it is silent and unseen. Yet even when we're asleep, the bats still chase moths and the oceans churn and lift, raising millions of animals up to the surface and then gently lowering them back down again. The circle of life is also the cycle of night, and in the following pages, we will explore these ideas in more detail.

⌦ Safety means many things; the participants on this Zodiac tour in the Solomon Islands are masked so they don't bring Covid to uncontaminated villages.

▲ A visitor's breath swirls blue in the humid air of a Japanese cave. Naturalist Paul Carter has completed one thousand cave visits, road surveys, and nighttime hikes.

The Milky Way spans all of Monument Valley in this multi-frame panorama.

The poet W. S. Merwin once said that on the last day of the world, he would plant a tree. I admire his optimism, but for me, I would find a nice high rock with a good view of the horizon—no camera this time, maybe not even my notebook, just a good view and an open mind. I would take a deep breath and try to become centered. And then, as the evening dusk turned into full night, I would lean back and watch silently as one by one each of the stars came out. Doing this is our oldest human experience, and still, after all these years, one of the best.

From Dotterels to Dragonflies: The Miracle of Migration

We start this chapter with birds, since they're the migratory group people know best. About half of the world's eleven thousand bird species migrate. Not all of this happens in the sky—some birds migrate by walking from higher elevations to lower, and some by swimming (say hello to the penguins)—and not all migrations are long-distance. But enough birds migrate five hundred or more miles that we can use their strategies as a way of accessing the rest of the natural world.

If birds migrate, it makes sense that one kind would out-migrate the rest. According to the internet, the prize for the world's best "mega-migrant" bird goes to the Arctic tern. This black-capped, white-winged bird is smaller than a seagull. It has a buoyant flight and hunts by plunge-diving for small fish. As part of its follow-the-sun lifecycle, this tern breeds in the high Arctic, and then as the days shorten it follows the sun to the bottom of the world, since in Antarctica, as the

◄ Arctic terns breed in places like Alaska, Iceland, and Norway, then migrate to the bottom of the world to spend the austral summer in Antarctica.

◄ The rufous hummingbird covers more distance in relation to its body size than any other migrant bird.

northern hemisphere goes into winter, the austral spring is starting to turn into the austral summer.

This long journey makes sense. By "wintering" in Antarctica,
the Arctic tern follows perpetual summer, which is really just
a way of following the food chain. In simplified terms, sunlight
means plankton and plankton means krill (and other organisms),
and these plankton-intermediaries mean food for surface fish,
and those fish grow into the just-right size category for terns.
And in the austral fall, as days shorten in the southern hemisphere, the terns fly north again, heading back to the tundra that
has now thawed in spring and become full of life.

During these journeys, the usual figure given is that the Arctic
tern flies twenty-two thousand miles a year—eleven thousand
miles down to the wintering grounds and the same distance on
the return trip. Other estimates double that total, since it's never
just straight-line travel for the terns—they need to forage and rest,
and they go around mountains or swing far out to sea to avoid
storms. A single tern might be flying fifty thousand miles per year.

Arctic terns can rest at sea, for example by perching on
driftwood (or trash). For a straight-line award, we have to look
at the bar-tailed godwit, a shorebird from Europe and Alaska.
Studies of radio-tracked birds found the Alaskan population
undergoes the longest known nonstop migration of any bird.
Apparently, these godwits take advantage of major pressure
ridges to catapult across the Pacific, covering distances of up
to seven thousand miles in nine days. They lose up to half
their body weight during these marathon non-stop migrations.
Shorebirds in general are all champion flyers, though sometimes they get pushed off-route by storms or a faulty internal
compass. The Eurasian dotterel is a small, tan plover that
breeds in the high Arctic and winters in North Africa. Yet a
few of these birds end up in the wrong places, many thousands
of miles from their intended destination. Dotterels have been
recorded in Hawaii, Bermuda, Japan, and islands in the North

Atlantic. I've seen one in California at Point Reyes National Seashore, a long way from its Norwegian breeding grounds.

Another way to measure migratory distance is to consider size. We all know that hummingbirds are small; the rufous humming-bird weighs just a few grams, about the same as a few paper clips. Being so diminutive, hummingbirds fly farther in relation to their size than other birds do. As Cornell's *All About Birds* website points out, "At just over three inches long, its roughly 3,900-mile movement (one-way) from Alaska to Mexico is equivalent to 78,470,000 body lengths. In comparison, the thirteen-inch-long Arctic tern's one-way flight of about 11,185 miles is only 51,430,000 body lengths." This means that in terms of lived experience, the rufous hummingbird "out-migrates" the Arctic tern.

Numerically, how many birds migrate? To get a clear picture, we need to think about all the mid-distance travelers—not the record-setting terns going south to Antarctica or even the Swainson's hawks headed from Alberta to the plains of Argentina, but the regular warblers and tanagers and flycatchers that are just going from Maine to Mexico or from Spain to Senegal.

The collective aggregate of migrating birds reaches staggering totals, in the billions. That is "billion" with a B—and as a reminder, a billion has nine zeroes—and it is one thousand units of a million. Adriaan Dokter of Cornell says that his team has "discovered that each autumn, an average of four billion birds move south from Canada into the United States. At the same time, another 4.7 billion birds leave the United States over the southern border, heading to the tropics." They winter in places like Costa Rica, and then as the days lengthen in spring and northern trees leaf back out, all these same birds fly back north again. Some have died over winter, killed by illness or old age or caught by hawks, but many survive. "In the spring, 3.5 billion birds cross back into the United States from points south," Dokter reports, "and 2.6 billion birds return to Canada across the northern US border."

A feral baboon on Gibraltar is unimpressed by raptor migration. "Seen one eagle, seen 'em all!"

Measurements in Europe show similar patterns. At a sort of natural bridge between Spain and Africa, Gibraltar, that limestone pinnacle guarding the Mediterranean, is a great platform for studying migrating raptors. According to the most recently published survey, each fall more than twenty-five thousand hawks, eagles, falcons, and vultures pass by a single observation station on Gibraltar. As they leave Europe and head south to sub-Saharan Africa, raptors are part of a movement of over two billion birds.

It's not just the birds themselves that are in transit. As one study points out, "Seasonal flows of extremely abundant migrants represent an enormous transfer of biomass, nutrients, propagules [stems, seeds, or spores that can grow into new plants], pathogens, and parasites, with effects on essential ecosystem services, processes, and, ultimately, ecosystem function." Or, to say it more directly: When you have that many birds, you have a heck of a lot of poo, and in that poo, you will have a lot of seeds being distributed to new areas . . . a lot of poo, but also diseases and hitchhiking creatures, such as feather lice. Not just birds migrate, but entire ecosystems.

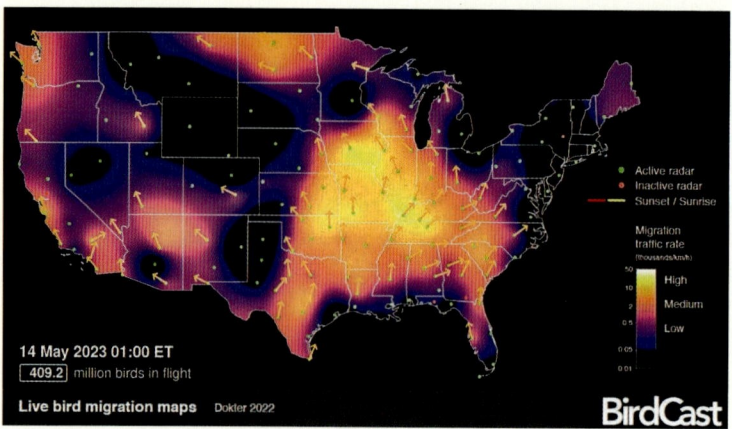

Active radar
Inactive radar
Sunset / Sunrise

Migration
traffic rate
(thousands/km/h)

50   High
10
2   Medium
0.5
          Low
0.05
0.01

14 May 2023 01:00 ET
409.2 million birds in flight

Live bird migration maps   Dokter 2022

BirdCast

This Bird-Cast online map shows in-progress bird migration at 1:00 a.m. on 14 May 2023, as captured by multiple radar stations at once.

Generally, birds migrate at night. It's safer, and (on average) winds are less turbulent. They can rest and feed during the day. Perhaps most important, they have access to a giant roadmap: the setting sun and the rising starscape. We will look at the specifics of nocturnal navigation in a moment.

Let's first review the science behind estimates of movement. We know what's overhead at night thanks to several techniques. In North America, birds in migration are flying in such dense numbers that they show up on weather-scanning radar. These observations are collated and made into maps by a project called BirdCast, and if you go to the BirdCast website, you can watch millions of birds moving hour by hour across the night-dark land. The colors the project uses for the graphics are simultaneously legible and attractive. (The images would make good posters or notecards.) During migration peaks, it might be said to be "raining birds," and so it makes sense to adapt machinery meant to sweep the skies, looking for rain and snow. Thunderstorm or bird cloud, the radar hums along, always looking, always ready.

Cornell Lab of Ornithology scientists analyzed data from 143 weather radar stations from 2013 to 2017 to provide the first large-scale counts of migratory bird activity across the United States. These radar results can be cross-verified against the records of birders logging into the tally system called eBird: If birdwatchers see ten sparrows and an oriole in an orchard, they note that, and all of the birders' sightings from across the country can be bundled together to show what a species like the field sparrow or Scott's oriole is doing on any given day during fall migration.

Other birders use acoustic monitoring to learn what is passing overhead at night, since migrating birds often use contact calls to keep in touch. Even a smartphone without a fancy microphone can be good enough to gather that kind of data, and software bundles can help sort out which species is which. Acoustic monitoring is accomplished with stationary devices, too, which are available to wildlife professionals under the acronym PAM, short for "passive acoustic monitoring." As our knowledge base grows and the PAM technology evolves, there even can be a chance to listen for everything at once. One recent article talks about recording a soundscape that synthesizes all biological, geophysical, and anthropogenic sounds in an eco-system. Bats, for example, echolocate at different frequencies from each other and often can be identified down to species by recording these calls and checking them against a computer readout. In the future it may be possible to listen for bats, frogs, crickets, and owls, all with the same tool.

Why Birds Migrate

Migrating is hard work, even for raptors like hawks that may ride on thermals in order to glide most of the way. And migration is risky, since you might get lost or not find enough food or get attacked by predators or be netted by poachers. You might fly into a building or miss an off-ramp and end up in Death Valley. Feathers wear out and need to be molted and regrown, and that takes energy; to be successful, a bird needs to learn the best

To make this nest in Alaska, the semipalmated plover parents flew thousands of miles. Can you spot the eggs? (Hint: The four eggs are in the exact center of the photo).

migration paths and be able to avoid hazards at both high and low altitudes.

As with the terns, birds are interested in resources, and so for migrating species, one must assume that to them the risk is worth the reward. In the tropics, the day is roughly the same all year long—about twelve hours. There is a lot of food, true, but there are other birds and animals trying to eat that same food, and there are a lot of specialized predators who want to eat you and your babies. In contrast, in the high arctic, the sun never sets in summer. The day is not twelve hours long—it is twenty or even twenty-four hours long. When there is more light, there is often more of other things, too: Insects are busy making use of the short summer just like birds are. Long days mean boom times for some kinds of prey, such as insects and lemmings; in migrating from the tropics to Alaska, you leave a relatively good feeding area in order to exploit the bounty offered by a much better feeding area. This bounty only lasts a month or two, and then birds shift back to more temperate areas to wait for the next cycle of abundance.

Yet it is not just food that causes birds to shift from wintering areas to breeding grounds. It can be another resource: nesting

A belted king-fisher leaves a nest site in Wisconsin—for this species, this is a hole worth fighting for.

sites. The belted kingfisher, for example, moves north because it is scouting for steep-sided streambanks into which to dig a nest burrow. These vertical embankments can be hard to find, and some of the best sites are under snow part of the year. Kingfishers don't mind the cold, but they need open water in order to catch fish. That means that a kingfisher may spend winter in a Louisiana bayou that has plenty of year-round prey, but to be able to mate and lay eggs, belted kingfishers need an embankment. To find that kind of location, they're willing to go a long way, even if it means flying from Baton Rouge to Juneau.

Kingfisher nests are located deep inside tunnels. It takes about a week for a belted kingfisher—usually the male—to dig a five-foot-long burrow, which they instinctually know to angle upward to keep out the rain. The main tunnel ends in a nest chamber the size of a small goldfish bowl. This burrow usually is in a riverbank near or over water, but it can also be in a sand pit, road cut, ditch, or quarry. Inside the nest, fish bones accumulate into a de facto layer of insulation; the poet Charles Olson once called this detrital layer the "rejectamenta."

All belted kingfishers have plumage that resembles a mohawk haircut and a stout, dagger-shaped bill. Both sexes are gray above, white below. Males just have one plain, slate-blue collar band over a white belly. The females are the "belted" ones, and they add a chestnut cummerbund across their white fronts in addition to the slate collar.

A special camera mount matched the rotation of the earth, allowing a full tapestry of stars to accumulate in one frame.

How Birds Navigate

Birds migrating at night navigate using a combination of cues, including looking at the stars, hearing ultra-low-frequency sound, smelling their way across the landscape, using polarized light, watching for familiar landmarks, and tuning in to Earth's magnetic field.

That's the short version.

The longer version starts with praising the immense beauty and complexity of the night sky, whose stars are more visible and more infinite than our streetlights and headlights usually let us experience. Birds are good at migrating—they have been doing it for thousands of years—and stars that to us may merely be decorative or a simplified set of mythological beasts, to an intelligent, alert animal can be a crystal-clear set of signposts. "Hey, everybody," the stars announce. "North is over here! And by the way, hurry up—it's almost four a.m.!"

In the above photograph from the Trona Pinnacles in the California Desert, we're looking out toward Death Valley National Park and the Mojave National Preserve. It's a moonless night and my campsite is dark—no campfires or strings of fairy lights. Even my headlamp is on a dim, all-red setting. My camera is set up on a tracker that matches the rotation of the earth, so I can do an all-night exposure and not have stars turn into white streaks.

This picture may seem like a science fiction scene or an AI version of an idealized nightscape, but the reality is that a good, dark, away-from-town night sky, as we look out into the thickest part of the Milky Way, is awash with stars and constellations. Unless you're stuck in a fogbank (or inside a zipped-shut tent), the night sky always has marker stars that can show you where you are, what time of night it is, and which direction to go to head north. On a night of a full moon, the light can be so bright it overpowers some of the subtle, more diminutive stars, but even so, directional clues still reveal themselves. That is because the moon rises in the east and sets in the west, so that, too, can be a kind of "follow the breadcrumbs" way to stay oriented.

Sound is another way to navigate. Birds can hear and see things we can't, and one thing they are aware of is ultra-low-frequency sound. They can create maps just from sound waves. Dr. Jonathan Hagstrum from the US Geological Survey explains: "Infrasound is very low-frequency sound. Pigeons can hear down to 0.05 hertz, while humans can only hear down to about twenty hertz. ... There are waves in the deep ocean that are constantly producing acoustic energy. And that acoustic energy travels through the earth as seismic energy, and then is reradiated at the landscape back into the atmosphere. ... Pigeons and other birds probably are listening to that reradiated infrasound, and using that to find their way home."

Human-produced static can block these soundwaves. If that happens, birds have other tricks. When you come back from a long trip, does your house smell like "home"? It probably smells more familiar and reassuring than most motels or cruise ships did on your holiday, or the rental cars you hired, or the restaurants you went to. Or to use a different thought experiment, imagine that you're in the most charming, quaint, postcard-perfect French village possible. If you were blindfolded,

These migrants have perished by hitting a lit building at night. Clockwise from top we have a blue jay, three black-throated blue warblers, a belted kingfisher, and an indigo bunting.

could you guess which way to walk to find the bakery? The smell of fresh bread could help you. Similarly, when you drive to the beach and get out of the car, suddenly the air smells "beachy," even without hearing the *scree scree* of seagulls.

Smell orients us to the world around us, and birds, such as seabirds called Scopoli's shearwaters, have been proven to "see" the world by smelling it. All parts of the ocean may look the same to us visually, but to a shearwater, they don't smell the same. Each piece of water differs from the next. Homing pigeons—and probably many other species—differentiate the world through smell, too. There is much we do not know about how other animals experience the world. They smell their way through the landscape in ways difficult for us to appreciate.

Light helps birds, yet it hurts them, too. Birds are not used to buildings standing a thousand feet tall in the middle of their flightpath, nor do they understand that lit bridges and buildings are not a kind of deviant moonlight, but urgent warnings of imminent collision. Sadly, migrating birds do hit buildings at night, and their bodies can be found on the sidewalk in the morning, as were the warblers and blue jay shown above. According to the National Audubon Society, "Every year, as many as one billion birds die from colliding with buildings, especially those with extensive glass surfaces. ... At night, when most birds migrate, lit-up buildings disorient and attract them, luring them not just off their migratory paths, but straight into collisions. These fatalities account for 2 to 9 percent of all birds in North America in any given year." Simple changes lessen this loss. You can add patterns to glass to help break up reflective surfaces, or building managers can do something as simple as turning off the lights at night during peak migration.

Yet birds need to use a special kind of light at the same time. You've experienced a version of this if you've ever worn polarized sunglasses. Professional car photographers screw polarizing filters onto their lenses to cut the glare off reflective metal. Botanical photographers do this, too, when working with shiny plant leaves. Physicists have known for over a century that right-from-the-source sunlight is unpolarized before entering the atmosphere, meaning its internal waves vibrate equally in all directions. Once sunlight approaches the surface, it changes. Light is polarized when the light reflects off a surface in a way that "bends" it. The degree of polarization is the percentage of reflected light that is bent, which depends on characteristics of the reflecting surface and the angle of incident light. Smooth, dark surfaces and low angles of reflection cause high degrees of polarization, including surfaces such as cars, oceans, roads, solar panels, and glass buildings. Landscape photographers know that putting a polarizing filter on a lens will help sharpen a scene and cut glare, but they also know that polarizers work best when at a 90-degree angle from the sun. Birds can "see" the polarization of light and use that awareness as another way of staying oriented.

Some day-flying migrants, like the European bee-eater, need landmarks more often than star-navigating night-fliers might.

Tourists may see the Hollywood sign, but to a bird, this is Mountain 1 (Griffith Park), followed by Mountain 2 (the Verdugo range), and then Mountain 3 (Angeles Crest). The Pacific is directly behind us in this view; a bird can hear the vibrations of the waves and see the polarized light in ways that a human cannot.

That species does indeed eat bees (though it usually knocks the stingers off first), and it has a breeding range that includes Spain, Denmark, and Italy, down to North Africa and across Eurasia to Mongolia and western China. Snow is not good habitat for bees, so this slim, kingfisher-size bird will migrate to Africa in winter. They migrate in groups of twenty to forty during daylight and will follow coasts, crossing the ocean at Gibraltar and other "chokepoints" that allow them to verify when they have moved from one zone to another. Bee-less expanses like the Sahara are overflown in a single push.

The European bee-eater is one bird that adapts well to a warming climate, since it is expanding its breeding range north. This species burrows into cliff faces to nest the same way the belted kingfisher does, and when they fly back from Africa, they show strong site fidelity, though if a river has flooded or changed course, the colony will change to a new location. All indications are that they see and remember landmarks with much better acuity than most humans achieve. No "Hey Siri, where's my car?" for them—they know where they are by paying focused attention to the landscape around them and can come back to the exact same cliff year after year.

The bee-eaters are the exception. While many birds watch for landmarks, most birds migrate at night. A final tool in the migrating bird's toolbox is the ability to tune into the Earth's magnetic field. One thing to understand about the idea of a magnetic "north" is that it does not match the tippy-top of the planet. Latitude and longitude are abstract concepts that mapmakers lay on top of an idealized sphere. The planet is not set up exactly that way in physical reality.

Before GPS units and other satellite-based technology, hikers used paper maps and analog compasses, and on the bottom of most topographic maps was a diagram indicating declination. You needed that to calibrate your compass. That is because true north might be the top of the map, but a compass points to *magnetic* north, which is currently under Ellesmere Island in northern Canada. It is near the North Pole but not *at* the North Pole. To get an accurate bearing using a paper map, one consulted the diagram printed on the map to set a dial on the compass to compensate for the "off-axis" position of magnetic north. How much one had to offset depended upon where one happened to be hiking; each map's printed adjustment notes differed from the next map's, based on location. This sounds more complicated than it was; in actual practice, for most hikes, nobody even had to bother with factoring for declination. Map north and compass north were close enough. Yet if you needed to know, you looked at the legend on the map and it was there.

Multiple studies have shown that birds have an innate sense of magnetic north, since they can "feel" (or "see") the Earth's magnetic field. Can humans sense it, too? Some people have a very good spatial memory, and if shown a horizon or other clues, can know approximately which direction north is. No studies have yet conclusively shown that humans can apprehend the Earth's magnetic field on their own, without secondary clues like using daylight or seeing mountain ranges.

Any single tool from the previous list might not be enough to help a half-ounce warbler get from Nebraska to Nicaragua safely, but, collectively, all the tools together make long-distance navigation a reliable behavior for all those billions of night-crossing birds. A few may go astray, but the majority won't. For them, the night sky is as inviting and legible as a turn-by-turn car navigation system is for us.

▲ Barn swallows winter in the tropics and migrate north to nest. This one is investigating the entrance to a motel in Arizona.

◀ From Paris to Panama City, this sign applies widely.

Bats and Migration

We turn now to a less known (and often mistakenly feared) group. Bats are the most diverse group of mammals on the planet, second only to rodents. Being so diverse, their approaches to movement vary. Around the world, many bat species are able to stay put. Fruit-eating bats in the tropics, for example, do not need to migrate far, since their climate and food supply are both stable. That is also true for non-fruit-eating tropical bats, such as fishing bats and even the sanguineous vampire bat, which makes scalpel-thin incisions on a sleeping cow's haunch and laps up the flowing blood. While there may be some local movement related to rainfall or food shortages, most tropical bats don't migrate far.

In contrast, insect-eaters in Europe and North America do need special strategies to survive a long, cold, bugless winter. They have two options: stay or go. Bats that overwinter in a cold climate will find a cave, well, bunker, or mineshaft and descend into a sleeplike torpor. While a bat hibernates, its body temperature drops and its metabolism slows; many caves are off-limits to humans then, since to awaken a bat midwinter means it will burn through its fat reserves too fast. As the expression goes, "let sleeping dogs lie" . . . and sleeping bats, too.

Bats that don't hibernate will have to go someplace warmer for winter. They will need to migrate like birds, and this is the bat cohort that concerns us now.

Mexican Free-Tailed Bats

One commonly seen bat in North America is the Mexican free-tailed bat. (Some sources call it the Brazilian free-tailed bat; both names apply to the same animal.) This species is a fast, acrobatic, high-flying bat that has been recorded feeding at ten thousand feet. They eat moths, including pests like cotton boll-worm moths and army cutworm moths, and they also eat beetles, blowflies, and flying ants. All of these are caught "on the wing." Some bats can glean prey from the ground, but this species is an aerialist, using echolocation to find flying insects.

Mexican free-tailed bats live in colonies, often in caves but also under bridges like the bats shown on the next page, which are circling under the I-80 freeway east of Davis, California. Soon they will move out over the nearby marshes and farms to feed. Nationally, by eating potential pests, bats save American farmers thirty billion dollars a year. That is both direct savings (crops saved) as well as the savings of not having to buy pesticide. If you like organic food, thank a bat.

Another large, easily viewed urban colony of bats lives under the Congress Avenue Bridge in Austin, Texas. That roost is home to 2.5 million free-tailed bats, and their emergence in summer is a great public event. People line up on the bridge itself, gather in the parks on each side, paddle in kayaks and canoes, or join dinner cruises to wait in the river below. The Congress Avenue Bridge bats will eat anywhere from five to seven *tons* of insects each night. Their prey base includes mosquitoes, so they collectively deserve an ovation.

Even in urban Los Angeles, a dozen freeway bridges have bats; the largest colony counted there so far (over the San Gabriel River, where the 605 freeway meets the 210) houses two thousand bats. Community science projects help monitor these sites, and exit counts document Mexican free-tailed bats and Yuma myotis bats, as well as the presence of other species (like hoary bats) that join in nightly foraging.

The bridge bats all roost inside the bridge's expansion joints. Because highway engineers know that bridges expand and contract during temperature changes, they design bridges to be ever-so-slightly "stretchy" rather than so brittle that they crack apart under stress. They achieve this stretchiness in part by incorporating expansion joints into the design. Bats can fit into a very narrow crevice, and they find that the concrete joints provide cool summer temperatures and safety from interference by snakes, humans, and other curious interlopers. A bridge or a cave or a vertical slit in a cliff face—it's all the same to them, so long as there are feeding areas nearby and nobody bothers them during the day.

▼ A swirl of Mexican free-tailed bats emerges from the I-80 freeway bridge near Davis, California.

▲ A volunteer uses a bat detector to make recordings as a long ribbon of bats flows from a freeway roost site.

As with other species mentioned in this book, the free-tailed bat's migration pattern is complex. Mexican free-tailed bats in parts of Nevada, Utah, Arizona, and southeastern California come together to migrate to Baja. Other bats in eastern Utah, southwestern Colorado, western New Mexico, and eastern Arizona travel through the western edge of the Sierra Madre Oriental

mountain range into wintering sites in the Mexican states of Jalisco, Sinaloa, and Sonora. Some bats that summer in Kansas, Oklahoma, eastern New Mexico, and Texas will migrate to south Texas first and then on to Mexico. And many populations of Mexican free-tailed bats that live in the tropics don't migrate at all: They already are where they want to be.

A bat's wing is really just a webbed hand—see the illustration on page 194—but wing shape varies depending on hunting strategies. Some bats have a long, narrow wing, good for tight turns and sudden acceleration, and others are designed for a slow, more patrolling flight. Some bats need to be more like a hummingbird; for example, nectar bats need to hover in order to feed.

The Mexican free-tailed bat has the fast kind of wings, and indeed, it can go sixty miles per hour, with a top speed of one hundred miles per hour. Peregrine falcons will still try to catch these bats as they exit their roost sites—nothing in the air is faster than a peregrine—and this bat species also has a high mortality around wind farms.

Mostly, though, this species thrives. It is the bat one sees at night exiting the main cave at Carlsbad Caverns National Park, and the largest known bat roost in the world (at Bracken Cave in Texas) is made up almost entirely of Mexican free-tailed bats. That site is home to fifteen million bats, whose nightly emergence is likened to a spiraling black tornado.

Interestingly, this species seems to be able to jam the radar of other bat species that would otherwise be competing for the same moths. Insect-hunting bats use echolocation to zero in on prey (discussed starting on page 197), and to be successful, a bat needs to get a clear "readout" of the returned call. A 2014 article confirmed that Mexican free-tailed bats had been detected emitting ultrasonic vocalizations that had the effect of jamming the echolocation of a rival species. Evolution works by trying to turn even small changes into advantages, under the general banner of "all is fair in love and war."

All bats are mammals, so they are warm to the touch; they have soft fur and small, sharp claws on their bottom toes. If you hold a live bat, as I have been lucky enough to do when working with researchers, they can be as light as a hummingbird. Face shape varies by species, but in general, bat faces remind people of dogs rather than of rats or flying mice. Indeed, there is even a species called the "dog-faced bat," and some kinds of bats look like pugs, with cute but "smooshed in" faces.

In turns out that each individual bat has its own personality. Researchers use fine mesh nets to capture flying bats. They measure the captured bats and sometimes band them or affix radio transmitters, then let them go. Just before the bats are released is a good time to take a portrait, like those seen throughout this book. When you hold a live bat for a photograph, their reactions vary. Some are curious about you, some are placid, some seem to enjoy being held, but a few are deeply vexed. They twist and try to bite, and if that won't work, they give you a scowl that lets you know they would gladly whup your butt, if only you gave them half a chance. Luckily, the Mexican free-tailed bat shown in the photograph here happens to have been one of the chill ones.

Dragonflies: The Unexpected Migrants

Dragonflies migrate—not all, but some. I remember how startled I was one night when, while looking for rattlesnakes and kangaroo rats in the Mojave Desert, I found a dragonfly resting directly on the ground. This was in a landscape not of streams and marshes, but creosote and Joshua trees. The little guy was not dead or injured—he was just in the middle of going from A to B and hadn't found an appropriate pond to stop over in. In the morning, after the sun warmed him up, he flew on.

All dragonflies have a tripart life cycle: egg, nymph, adult. Most live for two or three years, with most of that time spent underwater in the aquatic nymph stage. Nymphs look like six-legged, bug-eyed crickets and are voracious predators, gladly cannibalizing other nymphs as well as eating anything else their jaws latch onto, from worms to tadpoles to small fish. Dragonfly nymphs have rigid exoskeletons; they grow by shedding their exoskeletons in a molt and forming new, larger shells. Finally, two years after hatching, they are ready for one final molt. The mature nymph climbs out of the water, fastens onto a stem, and seemingly shuts all systems down. Out of this dead shell emerges a four-winged dragonfly, pale and fragile. Twelve hours later, once its wings harden and the colors fill in, the dragonfly is ready to fly off. Males dogfight other males over the best territories, fiercely protecting their chosen patch of pondwater. Females patrol as well, looking for good egg-laying areas. Mating pairs can be seen in flight over the water, looking strangely elongated as the male and female fly in tandem, joined in a midair embrace.

The female lays eggs soon after, then will repeat this cycle in other corners of a lake, pond, or stream. This is always a fresh-water activity; no dragonflies use ocean tidepools for egg laying.

If you want to determine which animal is the fiercest hunter of all, one number to think about is its success rate of attempted kills. Leopards and jaguars never achieve better than a 38 per-cent success rate, so for each animal they capture, two got away. Dragonflies, in contrast, have success rates between 90 and 97 percent. Part of that comes from aerial maneuvering: They can fly straight up, straight down, dart forward in a rush, or hover in place. Today no living dragonfly has a wingspan longer than six inches, and most are half that size. But before there were birds, before even the proto-birds called dinosaurs, the Earth was blessed with some immense dragonflies. They were among the first flying insects to evolve. The largest fossil dragonfly ever found, dating from 280 million years ago, sports a wingspan of two feet, four inches—a wider wingspan than a modern kestrel has. One theory to account for prehistoric dragonflies' large size is that Earth had high oxygen levels then.

Globe Skimmers Skim the Globe

World dragonfly migration has a spectacular "poster child": the globe skimmer, also known as the wandering glider. To understand its travels, we start with the monsoon cycle of South Asia. In summer, India heats up. Hot air rises, and other air—in this example, from the sea—flows in to replace it. This follows a predictable pattern, and in India's case, air from the southwest becomes hyper-saturated with moisture, and from May to October, this flow of wet air brings the monsoon rains to India.

In winter, as things cool down, the rivers of air are reversed. Wind then flows out of India, going from the northeast toward the southwest, back out across the ocean. If you drew an arrow to represent it, this prevailing wind would be pointed right at East Africa.

Under that arrow, as a stopping point along the way, are the Maldives. Because of the coral reefs, these atolls are full of dive resorts and honeymoon chalets; they also are a good base if you want to look for tropical species of whales and dolphins, as I did recently.

This is all postcard-perfect scenery; the white beaches and coral reefs are so fabulous they look photoshopped. Marine biologist Charles Anderson moved to the Maldives from England. As a year-round resident, he noticed things that the temporary visitors—the dive masters and the newlyweds—didn't think about. And one thing that he noticed was how every year, there were dragonflies. Not all the time, but for three months, there were thousands and thousands of dragonflies. He wondered how that could be. The fresh water in the Maldives is in a lens under the surface. As a villager (or a resort owner), you can dig a well to get down to it, but nowhere on the many islands and reefs and embankments do you have the kind of open pond water that dragonflies need in order to breed. So, logic said, they must be coming from somewhere else.

▸ This view shows a classic atoll, a Maldivian word that first entered the English language in 1625.

▸ Biologist Charles Anderson relaxes with a whale shark in Indonesia, asking it to please give his life preserver back.

It took several years, but Dr. Anderson worked it out. He made notes, talked to crews on research boats, called friends in other countries. Using their data on arrival and departure dates plus his own observations, he discovered a four-generation cycle.

It basically was this: The global skimmers breed in pools in India during the rains (generation 1); these were the ones he later saw in the Maldives. As India goes into the dry season, the dragonflies follow the rain by riding the wind to East Africa. This is the start of generation 2. As East Africa dries out, the skimmers next follow the rain to South Africa (generation 3). It's always a case of heat rising, which lifts the air, and new air flowing in to replace the missing air, and so for the dragonflies, it's a low energy cost to follow the dominant winds. In South Africa, as those rains end, the skimmers go back to East Africa (generation 4). From

there, they follow the monsoon flow back to India, when the rainy season is about to start all over again.

A two-inch insect is making a cumulative round trip of ten thousand miles over four successive generations. It is the only known transoceanic migration by any insect, and in total distance it is over twice that covered by a monarch butterfly.

One reason globe skimmers can endure such long-distance movements is that the convergence zones between high- and low-pressure systems are not cleanly vertical. Sometimes the wind flows meet at a slant, and that allows the dragonflies to look like they are flying against the wind, except they are not "against" the wind, they are above it, many thousands of feet up, using a counter-current. If this seems remarkable, that's because it is. One small insect, facing drought and desiccation with each successive season, has inadvertently figured out a way to extend the population's survival by migrating on monsoonal wind. If there is one single message of this book, it's that nature always finds a way, and globe skimmers are a perfect example.

And if there is a second message, it's that art, too, finds a way. We will not be reviewing all the times painters have captured a bird on the wing or a poet has written memorable lines about a bird, such as when Gerard Manley Hopkins praised a "dapple-dawn-drawn Falcon," calling it "morning's minion" (in this usage, the day's special favorite) and comparing the bird to lost royalty that outshines even the sunrise. We will, however, acknowledge one special moment of dragonfly art. Tebogo Monnakgotla, a Swedish composer, has written a violin concerto

A mallard swims through duckweed. During molt, until their wing feathers grow back in, ducks can't fly.

inspired by the globe skimmer's miraculous travels. It is being premiered in London as we go to press. Humans can be a bit of a mess sometimes, but some of us also know how to make—and how to listen attentively to—beautiful music. Why write a dragonfly concerto? Well, why not?

Things That Migrate (and Things That Don't)

Everybody knows that ducks can fly—and indeed ducks can, except when they can't. All birds need to replace worn feathers, and they do so by molting. Old feathers drop out and new ones grow in. The problem is, to fly, birds need most (though not all) of their wing feathers. Rather than dropping a few at a time, the way that gulls and hawks do, ducks just get it over with all at once. That means that at the end of summer ducks molt synchronously, which is a way of saying they lose their flight feathers in one intense push. This kind of molt sequence means that all ducks are briefly flightless until the new feathers come in. They can swim but not fly, so there's no migration happening during this period, no matter how strong the urge. They must wait until the molt is over, *then* they can go back to being the *Far Side* characters we know and love.

We finish by considering a mouse. One little-known North American rodent is the meadow jumping mouse. It lives in the lush grass beside mountain streams. This is a hard animal to spot because it only comes out at night, it prefers to stay hidden in dense grass, and it hibernates for much of the year.

Do jumping mice ever migrate, moving to another part of the mountains, or maybe down to a warmer, more snow-free meadow? It may seem like an odd question, but then nature continues to surprise us with oddities, from poisonous birds in New Guinea (if you catch one, don't lick its feathers) to the Australian platypus, whose fur glows under a UV flashlight. And other rodents do indeed migrate. As the Royal Society for the Protection of Birds (or RSPB, a British conservation group) points out, "Even moles have to migrate when their soil conditions change. If the soil becomes waterlogged during a wet spring, earthworms die. This deprives moles of their food—so they have to tunnel to new, drier ground. If the soil dries out during a hot summer, moles find it hard to dig their tunnels, and again they are forced to move on."

Another migrating animal is the Norway lemming; they are voles (small, short-tailed rodents) that, in good years with plenty of food, breed very fast. Each female has five to eight babies, and after only four weeks these babies can have their own litters. So many young are born that the lemming population explodes.

As food runs low, lemmings migrate, looking for better conditions down the road. They follow the most direct route and head for the highest point on the horizon, even if it means doing a mass "polar bear plunge" into an intervening lake.

In North America, biologists wanted to know if other rodents had short or long migrations. To find out where the jumping mice went and what they eat, researchers have tried fitting them with wee tracking collars, smaller around than a child's ring.

In the end, after tracking the movements of jumping mice for multiple summers in a row, researchers now know that the jumping mouse does not carry out long-distance relocation or movement cycles. Their home ranges are relatively narrow left-to-right but stretch out upstream and downstream to create an elongated rectangle. Theirs is a linear kingdom, very narrow

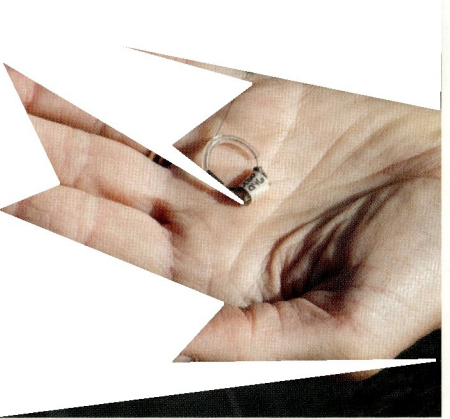

▲ A tracking collar is ready to be fitted onto a jumping mouse.

◤ This jumping mouse has its collar on and soon will be sharing data with forest managers and other researchers.

but very tall, even more stretched out than the country of Chile. It centers on the stream, though. Open-range cattle want water and visit these same sources of water and nice grass. That means that, if left undirected, range cattle can trample mouse burrows and wreck habitat. Land managers need to think about protecting riparian vegetation while still giving livestock access to drinking water. (Or we could just ban range cattle from public forests—but that is not a viable option right now.)

We can have both mice and cattle, but it takes the ingenuity and effort to build cattle-excluding fencing that protects the jumping mouse yet allows cattle access to the water.

Informed decisions start with good data, and good data can only be obtained one mouse at a time. In the forest night, doing their usual mousey things, the mice were helping create the data that would guide the plans. In a way, they were helping more than they could know.

Out there unnoticed and unseen is an advance scout for the next scientific breakthrough, in the form of a scurrying rodent pinging its tiny radio signal into the darkness, all night long.

Owls and Nightjars: Moonlight's Mystery Birds

R ogues, thieves, body snatchers, witches' familiars—that is how night birds have been portrayed in popular culture. And yet at the same time, owls are seen as symbols of wisdom, or they are spirit animals who can guide us to truth and insight, or, in the Harry Potter books, they are the sympathetic service animals that can also deliver the mail. We are not sure how we feel about them. William Blake, British poet and mystic, mocked people for being afraid of night birds. He said, "The Owl that calls upon the Night / Speaks the Unbeliever's fright." This implies that if you are strong in your faith, nothing going *hoo-hoo* in the dark should unsettle you.

That's easy for him to say. For most of us, the mysterious slamming of a door or the sound of a creaky set of stairs reminds us of horror movies. Things that we can hear but not see (including owls) make us uneasy. A barn owl's call can be downright spooky (one source describes their vocalizations as "screeches, wheezes, purrs, snores, twitters,

◀◀ A print from 1906 shows a hunchbacked witch with an owl, a snake, a spider, and a broom. This image could win a "most stereotypes in one place" award.

◀ The great horned owl looks like what we expect an owl to look like. It lives in deserts, forests, and even the middle of cities.

hisses, and yelps"), while the snowy owl matches that repertoire and adds raspy hoots as well, plus barks, grunts, screams, and whistles. The South American oilbird, a fruit-eating, echolocating, cave-dwelling nightjar (page 90), has the folk title of *diablotin*—little devil or imp. What we can't see or understand, we rush to demonize as "Other."

Even birders themselves are not always well-informed. Without accurate information, you end up with oddly hesitant ornithology. According to a recent book about all the nightjars of the world, a species called the Palawan frogmouth is "possibly not uncommon locally." What does that even mean? (I think the authors are saying that there might be some around in a forest somewhere, but nobody is quite sure.) To be "not uncommon" is apparently the best status a frogmouth can hope for. Maybe we should blame the name, which is not designed to make us want to seek them out and cuddle them. Children collect fluffy unicorns and stuffed pandas, not gape-jawed frogmouths.

A Central American pygmy-owl has caught a Salvin's spiny pocket mouse. Osbert Salvin was a Victorian naturalist who mostly studied birds.

The reality is that, mythology aside, owls are neither wise nor unwise. They do not ward off evil by hooting (as some stories claim), nor do owls generate evil by consorting with the devil. They are not a symbol of doom, human or otherwise. The only advance knowledge of death they possess has to do with the death of the rodent they plan to swoop down and eat.

We can make some generalizations about night birds' attributes. Broadly speaking, any bird featured in this chapter will be dull or cryptically colored. It will be able to see fine in daylight but even better at night. The birds featured in this section all have great hearing, and they all can fly (or at least glide) in near-perfect silence. In looking at night birds we will start with owls, since they are the most familiar and easily accessed. The "night bird" category also includes potoos and frogmouths, nightjars and nighthawks, and some final outliers like woodcocks that we will review last of all.

One easy way to start getting to know nocturnal birds is to find a sleeping owl during the day. That is easy to do, since they provide visual clues. A regular roost tree might be splashed with bird "whitewash," and there may be a few stray feathers under it, too. But the primary evidence comes from their meals. Some owls eat fish, but the typical owl eats small mammals. Their internal plumbing is such that they can't digest fur and bone, nor can they pass them out their hind ends. Owls get rid of this waste by coughing up gray wads about the size of your thumb; these packets of fur and bone are called "owl pellets." You can find them under roost trees, and they are good clues to help us answer two questions.

The first question, "Where are the owls," usually is answered, "Right overhead, sleeping on a branch just out of sight." Pellets accumulate on the ground directly under roosts, sometimes densely so. If it rains a lot or the park grass is overwatered, the pellets will dissolve after a few days. In that case, there are still usually some tiny bones worked deeper into the grass and a kind of gray soggy Kleenex remnant sitting on top. In contrast, if the pellets are deposited in a protected area, like inside a barn or under an overhanging cliff, they will accumulate for months, years, or even decades.

Once you find the pellets, you start birdwatching by looking straight up. Usually, the owner of the pellets is right above you, peering down.

The other question a pellet can help answer has to do with figuring out what the owls have been eating. In the photo on page 69, we know what the owl is about to have for dinner because we can see the menu item right in front of us. Of course, in the digestive cycle, what goes in must later come back out. So, a day or two after this photo was taken we could still identify this owl's prior meals because the tiny mouse skeletal bones would still be intact (albeit jumbled up), wrapped up inside the protective "hairball" of the regurgitated pellet.

In most climates, the pellet dries in the air and pulls apart easily. It is not gross or smelly. If you want to have a closer look, it's good to do the dissection over a piece of colored paper (which makes the small bones more easily visible and is better than a white background for photos) or a small pan or paper plate, so you can catch the lesser bits; a mouse's incisor is quite slender. You might want to wear gloves, and do wash your hands when you're done.

Owls and Owlets

Worldwide there are more than 250 owl species, and they fill out almost all habitats, from the Arctic to the equatorial rainforest to the bottom tip of Patagonia. A few have become extinct, usually on islands (Cuba, Mauritius), while other owls are expanding their range. The barn owl was not originally native to New Zealand, but now multiple pairs breed there. There is evidence that the first barn owls arriving in New Zealand got trapped inside the wheel wells of jumbo jets flying from Australia. Other New Zealand barn owls may have arrived by riding across the sea on cargo ships, or they were pets that escaped or were released and have since managed to thrive. Will barn owls keep expanding their range in New Zealand and become widespread and common? Nobody is sure.

All owls call, so if you don't find pellets, that is another way to find out where they are. Usually they vocalize before or just after dark. Smartphone apps can help identify the bird sounds you are hearing, and most field guides include phonetic transcriptions. According to *The Birds of Panama*, the rainforest pygmy-owl goes *pew pew pew*, but does so at a more leisurely cadence than the next closest species, the ferruginous pygmy-owl. Both differ from the look-alike Costa Rican pygmy-owl, which supposedly double-taps: *poop-poop, poop-poop*.

Good mimics like mockingbirds can further confuse things, as will be discussed at the end of this chapter. If you think you're hearing an owl, you're probably hearing an owl . . . unless of course you're not hearing an owl, but rather a non-owl bird imitating an owl.

Some owl species can hunt in complete darkness, using their dish-shaped faces and asymmetrical ears to locate rodents just by hearing them chew in the dark. Owl eyes are fixed in their sockets so owls turn their heads far to each side for peripheral vision—most owls can turn their heads 270 degrees. This ability to look backward makes it seem as if their heads can swivel in a complete circle—a kind of unworldly, even "possessed" behavior that adds to their cultural misperceptions.

The large number of owl species worldwide means that we can expect a lot of diversity. Indeed, owls are as varied as dogs

Small owls fit in small places. Here an elf owl makes a nest in a "just right" hole in a saguaro cactus.

are, so there are "Chihuahua owls" up to "Great Dane–size owls," and all the possibilities in between. That is one reason multiple species can inhabit the same forest, since each size of owl can specialize in a different target prey size. Some owls hunt during the day; some owls favor open habitat; and some owls are so poorly known that we're still not quite sure what their exact preferences are.

Since there are so many owl species, it can help to think of them in rough groupings or assemblages, sorted by size—small, medium, and large.

The "small owls" cohort includes species like the saw-whet owl of North America, the elf owl shown on the previous page in a saguaro cactus, the little owl of Europe, Central Asia, and North Africa, and the pearl-spotted owlet, whose African range is a sideways V going from Senegal to the Horn of Africa and then south through the savanna to South Africa. In most ecological models, there will be more plentiful lesser-size food items than larger things. Ecologist and author Paul Colinvaux acknowledges this in the title of his influential book *Why Big Fierce Animals are Rare*. In the typical landscape, there are more grasshoppers than there are mice and more mice than there are jackrabbits. Driving on a desert road at night, we might be more likely to see the jackrabbits (as they run in front of cars), but that doesn't mean the mice are not there and not more numerous overall.

That food base model means that small owls thrive because they can eat small things. In Africa, for example, the pearl-spotted owlet hunts during the day and yet also at night, and its diet includes insects (especially grasshoppers), plus small birds, bats, rodents, snails, and lizards—which is to say, it eats things that there are a lot of in almost all habitats. Like some of the other small owls, the pearl-spotted owlet has false "eye spots" on the back of its head, which make it seem like it is looking at you even when it isn't. These round eye spots may help discourage predators from sneaking up behind, or maybe the spots trick potential prey into thinking that the owl is looking somewhere other than right at them.

Little owls live in Europe and were introduced to England in the mid-nineteenth century. They look like the burrowing owl on page 82, except they nest in trees and old buildings, not ground-squirrel burrows. Sometimes in rural areas they were kept as household pets, where they companionably helped cottage dwellers by eating all the cockroaches. When they were first introduced from mainland Europe into England, they were rare and nobody was too bothered. Then as the little owls became more established, they began to spread. Some estate managers were worried the owls were going to eat all the pheasants and other profitable birds—birds they wanted to keep alive long enough for their guests to pay to shoot—and in a mistaken prejudice, advocated destroying the owls on sight. To disprove the myths, a woman named Alice Hibbert-Ware collected firsthand observational data, which she backed up by dissecting 2640 owl pellets. She conclusively showed that little owls eat insects, not hares and pheasants. Now attitudes have evolved, and there is great affection for this species in England and elsewhere.

The next jump in size takes us into common, visible species that many people know, including the eastern screech owl. This robin-size owl is found in the eastern United States, often in parks, woodlands, farms, and shelterbelts. Picture a map of the lower forty-eight states. If you imagine a vertical line between Canada and Mexico that runs through the middle of Texas (lining up Amarillo, Lubbock, and the Mexican state of Coahuila), and if you were to color in everything east of that line, you would have a good range map for the eastern screech owl. From the Great Plains to the Maritimes and from Canada to the Florida Keys, this generalist owl thrives anyplace there are trees (or trees plus human-supplied nest boxes). They even can be found right in New York City.

Tree-bark-colored and alert, a gray-phase eastern screech owl pauses at the entrance to its nest hole.

Midsize owls like the eastern screech owl eat the same things that most smaller owls do, but they can handle larger prey, too, such as squirrels, rabbits, voles, jays, and woodpeckers. If they have good luck, they may cache extra food for up to four days, but they eat most of their food shortly after capture. They nest in tree cavities (as shown above), reusing old woodpecker holes or abandoned squirrel dens or, should they find any, rot-opened tree cavities. Screech owls lay two to six eggs. The nestlings can be aggressive among themselves and sometimes will attack the weakest one—a grim practice called "siblicide."

Eastern screech owls come in two color phases, gray and red. Gray is the most common by about two to one. Perhaps this plasticity of color offers long-term options. For now, neither color phase outperforms the other, so both lineages survive and pass

◀ Many owls are midsize, like these crested owls.

on their color-favoring gene packets. A change in vegetation or other external factors could, many thousands of years from now, give one color phase an advantage, and in that case, maybe more of the population will trend in that direction until the population is uniformly red or gray.

One medium-size owl that I always like to see is the crested owl of Central and South America. What are sometimes called its ear tufts remind me of grandpa eyebrows every time I encounter them, giving this animal (at least to me) a kind of "Dr. Seuss face."

The crested owl can be found in dense, primary rainforest; secondary forest; and from sea level up to the cloud forest. It did not inherit a very tuneful song, though—its call is a cross between a wheezy "ghurrr" and a croak. (Several sources say this owl sounds froglike or toadlike, which is rarely a compliment. In *Phantom of the Opera*, the vain soprano, Carlotta, is cursed by having her voice changed into a frog's croak.)

A strange story about crested owl behavior comes from Brazil. Bat researchers had stretched out a mist net in order to capture and study forest bats. If you put a very fine mesh net across a stream, bats patrolling up and down the watercourse may fly into it. The thin string doesn't hurt the bats; they don't expect it to be there and get tangled up, but they normally receive no injury. The biologists are on hand to untangle the bats right away and to weigh and measure them, and sometimes they band the bats or affix radio transmitters to them. After the processing is complete, the bats are released and fly off to continue foraging. The whole process only takes a few moments.

Although the first rule of handling wildlife is "do no harm," some situations can't be anticipated. The researchers in Brazil had their nets up the same as always, but there was a problem. Two nights in a row, a crested owl heard (or saw?) the trapped bats and thought this was a good time to have a snack. Each night, the owl killed the bats that were caught in the net, hoping to eat them. But in reaching for a bat, the owl became entangled in the net itself. The researchers set the owl free and collected the deceased bat bodies to use as museum specimens, but they didn't know how common this "trophic interaction" was. Does this species of owl hunt streamside bats every night? If so, that makes sense. The species in the net, Seba's short-tailed bat, is one of the most common bats in this part of the rainforest. Or was the researchers' observation perhaps a one-off, meaning that the owl saw an opportunity and took it but probably won't hunt here again after the nets come down, since they will no longer be providing it with conveniently captive prey?

Both options are equally plausible. Add this to our running tally of questions filed under "more research is needed."

Moving up the scale, we get to the largest of the world owls, which in the case of a female Blakiston's fish owl, is a raptor that can weigh over ten pounds. (Female birds of prey are usually larger and heavier than males; they can utilize different prey bases than the males and, being more substantial, they can better defend the nest.)

Large owls engage people the most strongly. We generally like predators anyway—bears, tigers, sharks, crocodiles—and among these, the more fierce and potentially lethal something is, the more we like it. While small owls are cute, large owls demand respect, even fear. Maybe it's their binocular vision, with side-by-side eyes just like ours? Once, when I was showing one of my classes the campus great horned owl through a spotting scope, a student jumped back from the eyepiece. "It's looking at me!" she complained, afraid that the large brown owl wanted to attack her.

No, it had no such intent—it just was curious about the commotion below and perhaps a bit annoyed at having been awakened by our chattering. And that, as the pedagogy handbooks say, became a teachable moment. "What was the owl doing that was so frightening? Okay class, let's talk about our perceptions of the natural world."

The spectacled owl is a large owl of the New World tropics, found from Mexico to the Amazon Basin.

Out of so many great large owls to choose from, it's hard to focus on just one. Yet a perpetual crowd favorite is the spectacled owl. It is the size of a great horned owl and lives in Central America and the top half of South America. It is a handsome beast with a dark head, white eye crescents, and a pale cinnamon belly. It can eat things as large as a sloth but usually sticks to more opossum-size things, and it also likes to capture the agouti, a rabbit-size rodent we will meet on page 146. This owl will also eat jays and doves, as well as crabs, beetles, frogs, and iguanas.

Field-guide authors are not just guessing when they list some of these food items. Researchers know for sure that this owl can catch something as large as a sloth, because one ate a sloth that had been radio-collared—bringing a quick end to that particular study session. On a variation of "my dog ate my homework," I would hate to have to be the person who reports back to my major professor, "I am sorry, but I didn't get the sloth observations you asked for. A spectacled owl just ate our research subject."

Cornell University's "Birds of the World" website states that when it comes to the spectacled owl, there is "little information regarding abundance, population ecology, or behavior; there are no estimates of global population size or population trends." Note to animals: If you want to become well-known, don't adopt a nocturnal lifestyle. First, the early bird gets the worm, and second, the loud and obvious (and diurnal) creature gets the magazine spreads and the grant funding. Always better to be the elephant, not the elephant shrew. (Elephant shrews are real: see page 125.)

All owls are cool, but my personal favorite is the barred eagle-owl. It has the "eye wings" of the crested owl and the size and implied ferocity of the others in this widespread group, the eagle-owls. As a term, *eagle-owl* is a bit of a baggy fit; over a dozen owls have that term included in their common names. There is the pharaoh eagle-owl of North Africa and Arabia, the Cape eagle-owl of South Africa, the dusky eagle-owl of Asia, and so on. Of the fifteen birds with this same common name, half are related to the barred eagle-owl (genus *Ketupa*) and half are related to the European eagle-owl complex and are included in genus *Bubo*, the same group that includes the snowy owl and great horned owl.

Common names for animals rarely follow logic, and instead names reflect history, serendipity, tradition, and expediency. The great horned owl, familiar across the United States, could just as logically have been called the "North American eagle-owl." Latin names (the two-part genus and species pairing) provide some stability for thinking about animals, but for night birds, a lot of that taxonomy is still being sorted out. However many owls there are right now, and whatever their Latin and common names all are, if you wait ten years, there will have been a dozen updates. Nature is not static, but more important, neither is our knowledge base.

The eagle-owl photo here was taken in Borneo, so for me there was the added thrill of also sharing the same forest with orangutans and clouded leopards. Other subspecies of barred eagle-owl can be found on the islands of Sumatra, Java, and Bali. I want to see each one, just as an excuse to go to such interesting places. There is circumstantial evidence that this species is related to an African owl, Shelley's eagle-owl,

The barred eagle-owl is a "small" large owl, but is impressive even so. This one was photographed in Borneo.

and that it is related as well to India's spot-bellied eagle-owl. It seems to me that just to be complete, I should go to see those other owls, too.

Owls and Humans

Owls and humans have a complicated relationship. Owls appreciate the rats we bring everywhere with us, but some ignorant people still shoot owls, and as noted earlier, owls still have an unfortunate association with graveyards and Halloween.

One owl that is embedded in the human cultural landscape is the barn owl. It is a long-winged, open-country bird. As the name suggests, this owl does indeed roost in barns, as well as in other human structures like steeples and abandoned houses. By building and then abandoning these pseudo tree-cliffs, we serve an incidentally useful purpose for the owls. Yet we represent risk, too. Barn owls get hit by trucks and become roadkill with depressing regularity. In one all-day drive down Interstate 5 in California, I once counted eleven dead barn owls along the side of the road. In England, the data are even more grim. According to The Barn Owl Trust, "in a typical year, Britain's four thousand pairs of Barn Owls produce roughly twelve thousand young, and it is estimated that a staggering three thousand to five thousand of these are killed on roads."

The barn owl has a pale belly and a heart-shaped face.

In her book *The Wise Hours*, Miriam Darlington says that when it comes to humans and barn owls, "Our closeness has developed over time, like a marriage, but perhaps not an altogether happy one."

◄ A burrowing owl finds this artificial burrow "just right."

◄ A barn owl hunts over a controlled burn at a wildlife refuge.

One thing we can do to help species like barn owls is to put up artificial nest boxes. There is a shortage of dead trees with nice, watertight cavities in their trunks, so we can supplement existing nest holes by installing pole-mounted or tree-mounted nest boxes. These are usually made of wood and about the size of a cardboard moving box. The door hole can be round or square;

the specific size and features needed depend on whether you are putting the nest box outside, inside, or on a tree or pole. Not all boxes are the same; online discussion-forum threads show how to build owl nesting boxes so the babies don't fall out, how to weatherproof them, and how to keep out raccoons.

One not-so-obvious yet necessary feature of a nesting box is an access hatch, so that after breeding ends, the human caretakers can clean out accumulated pellets and debris. If the nest box floor in effect becomes raised by a thick layer of old bones, the chicks can reach the once-too-high, now-accessible nest hole and topple out. Cleaning out the pellets also provides show-and-tell objects for school tours.

Another species that does well with human-made burrows is the burrowing owl. This small, ground-based owl can hunt day or night. Usually it nests in (and chills out on top of) prairie dog colonies or ground squirrel complexes. Phoenix, Las Vegas, Tucson, Los Angeles—before these cities were cities, their landscapes were home to hundreds and hundreds of burrowing owls. With natural habitat altered or covered over, the owls need our help. Artificial burrows (like the one on the previous page) can provide that help.

A final owls-and-humans note involves fire. Although people may find the barn owl photo on the previous page alarming, the fiery scene is intentional rather than tragic. In it, the land managers at Aransas National Wildlife Refuge are carrying out a controlled burn for the general health of the marsh. By helping keep salt meadows open (and not letting them become filled in with shrubs and trees), they do the barn owl a long-term favor. Controlled burns do owls a favor in the short-term, too: The small, quickly burning fire will flush prey into the open, and the owl is exploiting the situation fully.

Tawny frog-mouths are crow-size birds that live in Australia. In this view, the bird is looking backward. The beak is above the right shoulder.

Frogmouths and Potoos

The strange and mesmerizing frogmouth complex is made up of sixteen species found in Australia and Asia. A superficially similar group, the potoos, lives in the Americas and is comprised of a total of seven species. They look like each other, but this is an example of convergent evolution; the two groups are not closely related.

Individual frogmouths and potoos look like mossy tree stumps and have large mouths (or "gapes") and big eyes—or at least they have big eyes at night, when hunting. During the day, when the birds are sitting still and wishing you would just go away, their eyes squinch down to the teensiest of slits, as their faces turn into an exact replica of tree bark. Even when you're looking right at one, it can be hard to spot it in the dim light of the forest—they really do look just like dead trees.

These are all sit-and-wait hunters, exploiting edges and clearings and (on average) sallying up to take passing insects, though some of the larger frogmouths can and do eat frogs. If you want to sound like the ultimate science nerd, and if you know that

A Bornean frogmouth waits patiently for a passing insect. Females, like this one, are more reddish brown; males are a plainer, grayish brown.

antelopes are called bovids and that seagulls are called larids, then you might want to show off by using the word *podargid* to refer to frogmouths, from the family Podargidae. The word *potoo* comes from a transcription of the bird's call. In Africa, a similar word, *potto*, refers not to a night bird but to a small, nocturnal primate. These two terms get mixed up sometimes; see page 223 for more on pottos.

On foot or by jeep, if you go into a tropical forest after dark and scan with a spotlight, you might see an animal whose presence is given away by the bright glow of their eyeshine. With potoos especially, their eyes reflect light from a long distance away and can be so bright that they appear not to be an animal sitting on a fencepost or tree, but instead perhaps a lighthouse or landing airplane way *past* the fencepost or tree, glimpsed through branches.

Animal spotters debate the ethics of using flashlights to study animals at night. Never harry any animal, especially if it is carrying young. But as a general defense of all-night animal spotters, most animals have to cope with lightning as part of their natural environment, so a flashlight is not without precedent. The

A northern potoo in Belize practices its "you can't see me" pose.

Bornean frogmouth shown on the previous page was indifferent to the people in the jeep below its perch, and it did not react (or move away) when we used a flashlight to help focus the camera to get the shot. Be ethical and mindful *always*, but animals are durable—if they were not, they probably would not have lasted all these many millions of years.

These birds are so strange, it's hard to know how best to describe them in words. They just don't fit into our usual categories. To one early reader of this manuscript, the frogmouth on page 84 looked like "a moth that had met the blue faerie." There probably was not magic involved, but given that we can't see all the steps of evolution, that origin story is as good as any.

The take-home message for all of these species is how well cryptic coloration can work. We might expect an insect no longer than your little finger to be able to mimic a dry stick. These birds are much, much larger than an insect. The great potoo, *Nyctibius grandis*, has a two-foot wingspan. They may as well all be wearing invisibility cloaks, since their swirls blend in with bark so well. There was one potoo in Ecuador that I must have been looking straight at for ten minutes—with the guide patiently explaining in Spanish, "it's there, *right there*, on the dry stick"— before I finally said, "Oh my gosh!" and saw the bird. What I had thought was a piece of dead tree was a potoo looking straight up, eyes closed and beak held vertical to the sky.

I asked the guide how he had ever seen it to begin with.

He laughed and confessed, "Oh that one, he is always sitting there. I show the tourists every day, all the days of the week."

Nightjars, Pauraques, Poorwills . . . and One Oilbird

Cryptic as potoos but usually posed more horizontally, nightjars sally high after passing moths, and they usually do so from a roadcut, a pasture, a rocky hillside, or a storm-cleared section of a tropical forest. During the day, nightjars rest on branches or sleep directly on the ground, blending in with the gravel or dry leaves around them. There are ninety-six species of nightjar worldwide including the poorwills, pauraques, and the nighthawks, which we will meet in a later section.

◄ This rufous-cheeked nightjar in South Africa shows all the expected nightjar features: large eyes, cryptic colors, and a ground-hugging posture.

◄ Ethiopia's Nechisar nightjar is known only from a single wing removed from a decomposing corpse. Nothing more is known about this mysterious bird.

Nightjars, like owls, are another feared and misunderstood group, and in past years they were given folk names like *goat-sucker*, *lich fowle* (which means "corpse bird"), and *goat-chaffer*. In Norwegian, the name translates as "night raven." Even the regular "nightjar" name has a bit of discord built into it, since only unwanted intrusions, like a bad dream or a wrong number at two a.m., will jar us awake in the middle of the night. Once we move past the name, we meet an interesting creature. In their own way, they are very attractive; the vermiculation on the backs and wings takes the idea of camouflage to artistic heights.

The common names nightjar, poorwill, and pauraque all describe the same kind of night bird. This happens often with birds; "heron" and "egret" are part of the same group, too—two names for what is basically the same floorplan. "What's in a name?" Juliet asks. Often, with birds, not much.

No matter what we call it, the archetypal nightjar is a swirly mix of gray, tan, and brown. In flight it has some white on the tip of the wing or edge of the tail; it has a white throat band (or a buff collar); it vocalizes with a churring trill (or else the alarming "whip, whip, whip poor Will"); it nests on the ground; it hunts by sight, not sound, even on the darkest nights; it may be most closely related to swifts, hummingbirds, and frogmouths (or then again, it may not); it has a small bill and a cavernous mouth (or "gape") surrounded by facial bristles to help detect prey (and maybe protect eyes); and, for some species, the males have long wing pennants or tail streamers for spectacular displays in courtship. To understand a nightjar, it can help to picture them as something the color of a screech owl but shaped like a resting-on-the-ground, about-to-be-hiked American football. The nightjar in question, whether in the desert or forest, waits on the ground and then flies up with a sudden white wing flash to snatch a passing grasshopper.

Few of the ninety-six nightjars are well-known, and in some parts of the world, scientists are not even sure which ones have

gone extinct. The Jamaican poorwill has not been definitively recorded since 1860, while the New Caledonian nightjar is known only from a single specimen collected in 1939. It has not been seen since. Is it still out there? Please call collect when you find out. Something called the Nechisar nightjar is known only from a single wing found in 1990 in southern Ethiopia. Nobody knows what the rest of the animal looks like.

The Indian nightjar's call sounds like a stone skittering across the ice of a frozen pond.

An especially odd nightjar is the South American oilbird, so named since the bird's plump young make good torches. (Oh humans, *really?*) It is red-brown with small white spots. The pinkish bill is strongly hooked. It eats fruit, it echolocates, and it breeds and roosts in colonies in caves or steep, dark, overhung ravines. One guidebook says that it "makes a variety of strange clicking and shrieking noises"; a different source describes the vocal range as "a variety of screams, snarls, and snoring sounds." (I would hate to hear what these authors would say about my own sleep habits.) It is very strange and awaits further investigation.

Even some birds from the United States are still mysteries. One common night bird of the borderlands is the Mexican whip-poor-will. According to Cornell University, this bird's "biology is essentially unstudied, largely due to its nocturnal activity and cryptic behavior. Indeed, it is among the least-studied breeding species in North America."

One nightjar that we know a bit more than usual about is the Indian nightjar, shown opposite. It occurs from Pakistan through all of India and Sri Lanka and in an arc across Myanmar, Thailand, and Vietnam. Depending on which source you trust, its call sounds either like a stone skipping across an icy pond or a ping pong ball bouncing down a marble staircase.

We know that this species usually eats moths, beetles, grasshoppers, crickets, locusts, flies, and ants. So far, everything on that list is just as expected. But during non-breeding season, the Indian nightjar reportedly eats *Euphorbia* flowers too, and also mice—in particular, the Indian fawn-colored mouse. Do other nightjars do this? Nobody knows.

Animals (just like people) will go to extreme measures to attract a mate. Among nightjars several species deserve special notice. In the Andes, the male lyre-tailed nightjar has tail streamers that are twice the length of its body. We're used to talking about a peacock's tail, which isn't really the tail but a wedge of modified back feathers. It looks flashy but is not very long compared to the rest of the bird. For the lyre-tailed nightjar, in relation to our own size, it's as if a five-foot-tall woman was trying to walk around in a bridal dress with a ten-foot train, and not just walk, but fly, and not just fly, but fly with that kind of dress not just once only, but during all the days of the year. Going to work, going to the gym—always in a bridal dress with a huge, long train. Size of tail probably indicates overall health and the ability to forage well and the ability to find a good day roost without breaking feather shafts. Long tail feathers could also help during tight turns and aerial maneuvers. Mostly, though, it comes down

to Darwinian display: You can't pass on your genes unless you raise young successfully, and so the "longer tail" genes will win out and continue to the next generation.

A lesser nighthawk swoops low, drinking on the wing.

In Africa, the standard-winged nightjar has a display aid that even Wikipedia calls bizarre. During mating season, the male bird has an elongated shaft at the end of each wing that ends in an extravagant plume. This is the "standard," in the sense of meaning a pennant or flag. If it were a person, it would be as if a cheerleader were standing on the sidelines with a giant, oversize feather duster in each hand. During a display flight, these standards rotate upright and start to flutter, sort of like a castaway on a raft waving a shirt, "Over here! Over here!"

We hope that female nightjars are impressed; the male has gone to a lot of trouble to maintain these extensions and haul them around all day.

Nighthawks

Nighthawks are a subgroup of nightjars that tolerate cities better than most birds (and can even nest on rooftops). Compared to the typical nightjar, nighthawks have longer wings, brighter wing flashes, and a more buoyant flight style. One expects to see them in open country, and two kinds, the lesser nighthawk and the common nighthawk, are widely known in the United States. Common nighthawks call *peent, peent* as they hunt over wetter landscapes, including meadows and lakes, evening baseball games, and even well-lit highway billboards. Lesser nighthawks are desert birds. On hot summer nights you can see them coursing over the Las Vegas strip, attracted to the moths that are drawn to the lights.

Experts can tell the lesser nighthawk from the common nighthawk based on wing patterns, but many people don't bother. These two species are very close look-alikes, more easily separated by voice and range. Both species have obvious white wing flashes (as shown in the photo on the previous page) and both come out earlier in twilight than most other nightjars. Both have a long-winged, slightly rocking flight.

As an example of how night animals don't get documented because nobody is paying any particular attention, consider this story of an urban bat survey I was doing one night under a major Los Angeles freeway. The road crossed a river here and the bats roosted under the freeway, inside the bridge; the volunteers all had clickers to count bats at dusk as they came out. There were several other volunteers present that evening, and of course there were thousands and thousands of people an hour rushing by inside their cars on the freeway. A bike path paralleled the river where we were doing our survey, and fit, determined riders on six-thousand-dollar bikes pounded downstream at thirty miles an hour, their headlamps blurring past in a streak of white.

My handheld bat detector helped confirm who was who in the mix above us—Mexican free-tailed bats, Yuma myotis bats, canyon bats, and hoary bats. As the sky turned from orange to deep blue, a new silhouette alerted me that a nighthawk had come down the canyon to join the feeding melee. Searching with my headlamp I could pick out the white wing crescents as it veered and jinked. No other team members noticed it; they were intent on counting the bats.

We see the nature that we allow ourselves to see, and we see what we are aware of as a possibility in the first place. As the bat team focused on their specific tasks, small toads (toadlets, really) scurried underfoot; I tried to be careful not to step on any. I also tried to categorize the mulchy dampness of the river's many smells. Yet at the same time I am sure I was oblivious to many moths that were coming out, and I failed to notice what time the dragonflies went to roost. A great horned owl called far away in a city park—I almost didn't hear it over the busy hum of my own thoughts. The night is rich and well-populated, which we soon discover if we just open ourselves to it.

Other Night Birds

Other bird groups besides owls are nocturnal. For example, there are seven species of night heron worldwide, and in New Zealand, the famous kiwi is a round-bodied, worm-eating, shaggy-feathered night bird of the forest floor.

One nocturnal bird that Shakespeare fans may remember is the woodcock. Both halves of the name can bring up sniggering sexual references, as the Bard knew full well. He was never above a bawdy joke to provoke the crowd. In one scene, preoccupied with his daughter's chastity, the court advisor Polonius accuses Hamlet of using sweet words to seduce Ophelia—the prince's words, Polonius says, are like "springes [traps] to catch woodcocks." Hamlet has in fact been acting honorably, but Polonius has a one-track mind, and even his own comparison is inadvertently risqué. This passage reminds us how much nature lore permeated premodern society and shows that in Shakespeare's time, even the nobility ate wild-caught table fare.

A woodcock is a large, forest-dwelling shorebird, cryptically colored, and in both American and European species, the male makes roding flights—that is, it performs nocturnal courtship flights by ascending high in the air and then spiraling back down while calling. Under the heading of "nature is all around us," I saw my first woodcock late at night in downtown Boston, in a snowstorm, outside of the Boston Public Library. The woodcock photo on page 97 was taken in Central Park, New York City. Usually one goes to the countryside to find these birds; preferred woodcock habitat is shrubby forests, old fields, marshy edgelands, bogs, and powerline cuts that transect mixed forests.

Woodcocks feed by probing into soil, hoping for earthworms and also interested in ants, beetles, snails, and millipedes. As if listening to a funky beat on earbuds, woodcocks rock forward and back, silently dancing across the ground. This "two steps forward, one step back" shuffle probably helps flush their prey, which they can detect by sound. Hey, worms, you can run but you cannot hide. Woodcocks also eat leeches and, less commonly, pigweed and ferns.

The male's sky show is diagrammed on the next page (based on an illustration from sibleyguides.com). The data-aggregating site eBird describes how, when the time is right, a calling male "eventually launches into the air, flying rapidly upward in spiraling flight while producing a twittering wing sound. At the apex of his flight, he suddenly goes into free fall, floating back to the ground like a falling leaf while giving a distinctive chirping call. Then, just before reaching the ground, the male will resume

The flight display of American Woodcock

Besides being namechecked in Shakespeare, the woodcock lives on in other artistic expressions. John James Audubon and Louis Agassiz Fuertes both did fabulously good woodcock portraits, and woodcocks have also appeared on dozens of postage stamps worldwide.

The crepuscular woodcock probes damp soil for earthworms.

The most insightful comment comes from the great conservationist (and author of *A Sand County Almanac*) Aldo Leopold. In his lifetime, he saw the end of "shotgun ornithology," the now-concluded era when birdwatching centered on shooting specimens or raiding nests for eggs. His writings helped many thoughtful naturalists make that pivot. For Leopold, beauty was always reason enough for conservation, and he asserted that the woodcock's intricate displays were a "refutation of the theory that the utility of a game bird is to serve as a target, or to pose gracefully on a slice of toast."

Here Come the Moths! Plant Pollination after Dark

I f you've ever decided to fly out of a small, regional airport instead of the big city mega-hub, you can appreciate that crowded skies are rarely friendly skies. From parking to toilets to the line for coffee, overly busy airports make passengers compete for limited resources. You can almost taste the stress in the air, hanging like invisible cigarette smoke.

Plants face a similar dilemma. If they want to be pollinated, blooming in the daytime is complicated. On the one hand, they have lots of potential suitors: There are 350 species of hummingbirds worldwide, plus bees, plus butterflies, plus flies, plus the wind. Plants can even release pollen into a stream or pond and try to connect with destiny that way, relying on something called surface hydrophily.

Yet on the other hand, despite these plentiful options, every other plant is trying to flag down the same taxi, hoping to be pollinated, too.

A sphinx moth visits a desert willow at dusk. The long "wire" is the tongue.

No matter how red a plant's flowers may be or how perfumed their scent, other plants are competing for the same services.

Some plants respond by blooming at night. From saguaro cacti to mango trees to the blue agave cultivated for tequila, many plants have developed partnerships with nocturnal pollinators. No hummingbirds for them; these plants bloom only (or mostly only) at night, and they are visited by bats, moths, or even frogs, depending on the plant, the habitat, and the time of year. For a desert plant producing a large bloom, that represents a considerable water investment, in terms of nectar and growing the blossom itself. Opening at night is a water-wise choice.

Behind these floral displays is a basic biological imperative: The goal of every organism is to create successful offspring. The question is, what is the best way to do that? Some plants can reproduce asexually (through budding, fragmentation, fission, spore formation, or propagation), but most plants benefit if pollen from plant A comingles with pollen from plant B. Blending traits from two parent strains helps maintain genetic diversity, which in turn helps plants achieve generational continuity.

One example of a night-pollinated plant is the iconic Joshua tree of the Mojave Desert. Most people know this plant from movies and the title of the 1987 U2 album; it is a tall, branching yucca found at higher elevations than so-called low desert plants. If there has been a wet winter, in spring the Joshua tree produces big, showy sprays of creamy white blossoms. By fall, seedpods develop, fall off, and break open, spreading seeds across the ground. A new yucca plant is potentially on its way.

Yet this process depends on the springtime arrival of a tiny nocturnal insect. That is because Joshua trees can only be pollinated by one visitor, the yucca moth. This little guy is eensy—barely the size of a grain of rice. Yucca moths mate inside the flowers of Joshua trees. The female uses her adroit jaws to collect pollen from the male part of the flower, then flies to a new tree to deposit the pollen on the next flower. This helps ensure

▸ Desert icons, Joshua trees stand silhouetted by a vivid sunset.

▸ Researchers set up a moth trap inside this Joshua tree blossom.

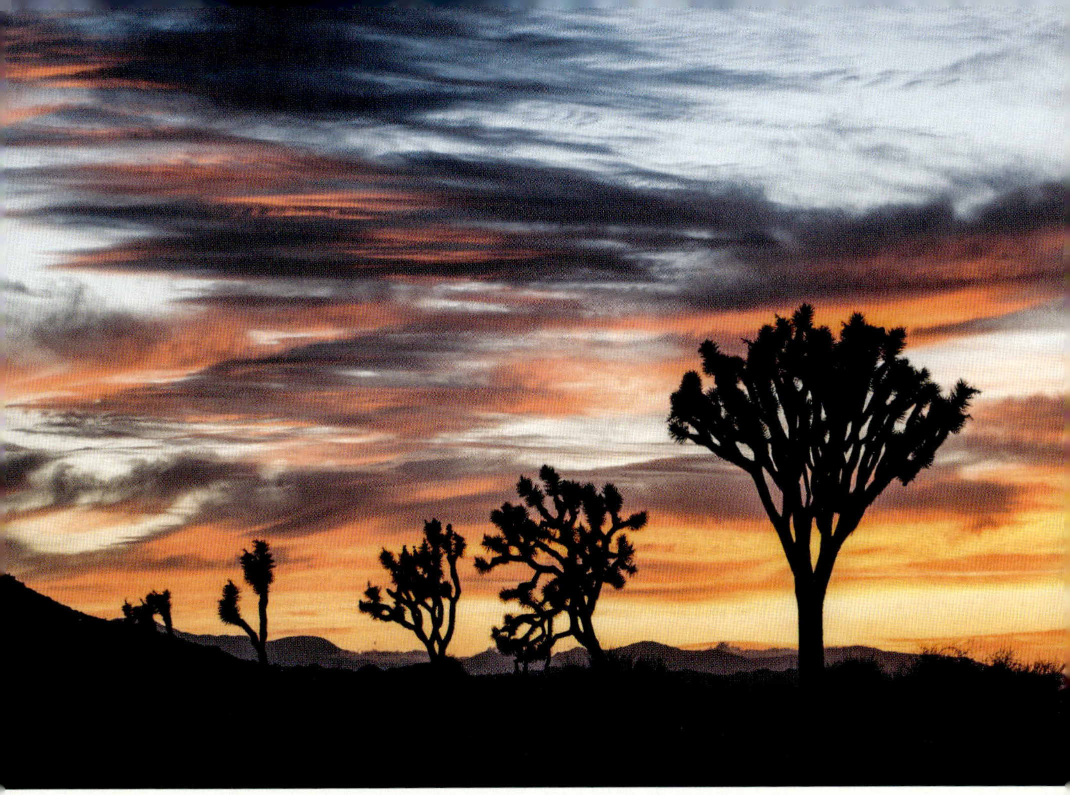

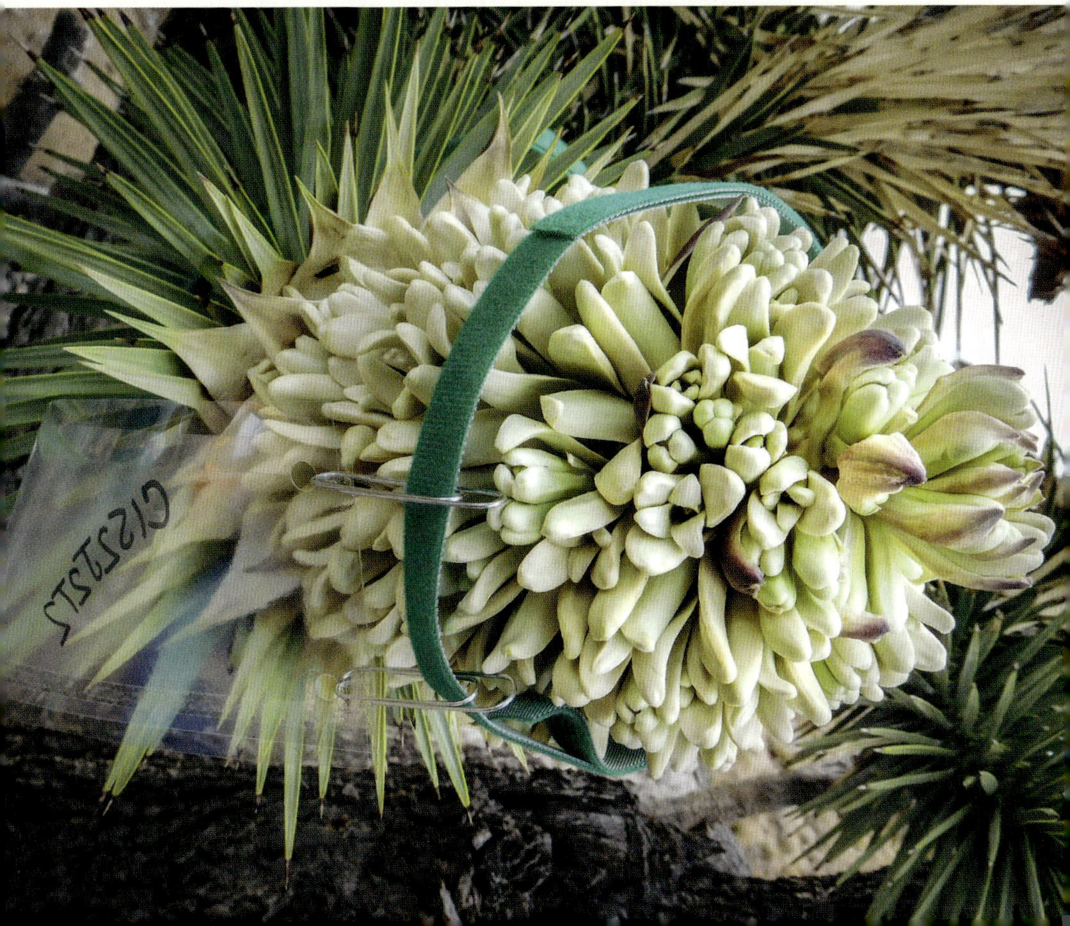

the plant's fertility. The female moth lays her eggs by the part of the flower that will become the seeds. The caterpillars hatch and eat some of the seeds, then work their way down to the ground to make cocoons. The remaining seeds end up as part of the general seedbank in the soil, ready to help grow new Joshua trees in years to come.

Moth-a-Palooza

Some flowers swing both ways: They are visited by bees and other pollinators during the day and by moths at night. On average, moths make better pollinators than bees, since adult moths tend to move longer distances between patches of plants than bees, which usually forage in a narrow radius around a specific hive. Further, moths have fine hairs on their bodies that help trap pollen grains. They can carry more pollen per load than other insects.

For plants involved in the bloom-at-night trade, the Joshua tree's "one moth, one host" model is not the norm. It *can* be, but most moths typically visit a variety of plant species seeking nectar. Brown-eyed evening primrose, for example, relies on moths for pollination. It opens near sunset, blooms all night, then withers in the morning. Datura, a showy white plant that grows in disturbed soil like landslides and roadsides, can bloom during the day (and if so, it will be visited by bees), but is primarily a night bloomer advertising for moths. The twenty-inch-tall cranefly orchid lives in woodlands and along ditch edges. Because it lives in shaded areas, it reverses the normal botanical calendar. In late fall or early winter, when most of the deciduous canopy has dropped its leaves, each orchid will produce a single green leaf, called a hibernal leaf because it is present only during winter, when many other plants are dormant. By late spring, the window for photosynthesizing closes as the tree canopy fills

in with new leaves. The cranefly orchid's leaves die back then. No leaves are left by the time the plant starts to bloom in high summer. The plant sends up a thin stalk with up to forty delicate, attractive flowers—each one ready for a visiting moth. No flies are involved in its pollination; cranefly orchids are named for their flowers' resemblance to crane flies.

Hawk moths are among the largest and most easily seen night-pollinating moths. Plants that attract hawk moths are called *sphingophilous* plants—be sure to use that word at a party—which means that their flowers evolved to have deeply sequestered nectar accessible only to visitors like humming-birds (long beaks, long tongues) and hawk moths (no beaks but even longer tongues than hummingbirds have).

Moths differ from butterflies structurally. One thing to look for are the antennae sticking out past the face. A butterfly's antennae are club-shaped with a long shaft and a bulb at the end. A moth's antennae are feathery or saw-edged—they look more like an old-fashioned comb. If you can't tell just by looking, try taking a phone shot and then zoom in. Butterflies are larger (often, not always) and more colorful (often, not always), and butterflies hold their wings vertically at rest, making the shape of an A-frame chalet. Moths are typically smaller with drabber wings, and they more often rest with wings held open flat. Butterflies are primarily diurnal, flying in the daytime. Moths are generally nocturnal. However, there are moths that are diurnal and there are butterflies that are crepuscular, flying at dusk and dawn. And it is also the case that not all moths pollinate; just the same as with butterflies, some species do not feed at all as adults, and others that do feed will seek out rotting fruit or tree sap, not flowers. Moths have even been reported to lick the tears from sleeping birds. Bees and flies do it, too, and this kind of feeding has a name, "lachryphagy."

One overall truth about moths is that they deserve more respect than society usually grants them. From boxing similes like "float like a butterfly, sting like a bee" to elaborate tattoos in intimate places, butterflies get all the attention. Meanwhile, if there is a moth mentioned in popular culture, it's always connected to something bad, like the expression "a moth-eaten sweater" for a ratty piece of clothing. "Drawn like a moth to a flame" is a common expression about a fatal attraction; death's-head hawk moth is a real species, but there's never an equivalent entry for the death butterfly. A navy's "mothball fleet" consists of surplus ships, quietly rusting at anchor until another war makes them desirable again.

These negative associations are not fair; if viewed more objectively, moths have a lot to offer. One moth that is called the miller moth (or the army cutworm moth) comes out in summer. They migrate to mountain regions each summer to feed at night on the nectar of alpine and subalpine flowers, and during the day they seek shelter under rocks. In the Rockies, grizzly bears feed almost exclusively on moths for up to three months straight. We think of bears wanting to take down an elk (or at least a honeybee hive), but army cutworm moths are a preferred source of nutrition for many grizzly bears in the Yellowstone ecosystem.

The bullseye moth is native to South America and could compete with any butterfly in a beauty contest.

During a single day in August 1991, fifty-one bears were counted feeding at just four moth aggregation sites.

The bullseye moth is a nocturnal species from the Caribbean and South America. Perhaps because it looks so much like a butterfly, it does have a large fan base (at least for a moth). Some insect-lovers opt to raise these moths in home terrariums, while other hobbyists paint their portraits as craft-for-sale watercolors. Photographers delight in capturing bullseye moths in images. You can also buy specimens of this species premounted in shadow boxes as a kind of morbid wall decoration, but we recommend appreciating living animals, not mounted ones. Alive or dead, the bullseye moth can appear in a variety of hues. In duskier versions (with darker wings than those of the moth shown above), the bright orange wing spots glow as if lit from within.

Another interesting night pollinator is the Virginia creeper sphinx moth. It looks like dried leaves—or should we say that dried leaves look like Virginia creeper moths? These moths can be found in the eastern half of North America, including in southern Canada and in the United States from Maine south to Florida. They fill in the map westward to North Dakota, Nebraska, New Mexico, Oklahoma, and Texas and can also be found in Mexico. This insect's common name comes from a host

The Virginia creeper sphinx is named for a vine, not the US state near Maryland. In this photo, the hind wing pokes out from under the larger, more swept-back top wing.

plant for its caterpillars, the vine known as Virginia creeper, which has small blue berries and can grow to be forty feet long. The vine's leaves often turn red in early fall. Virginia creeper looks a lot like the rash-causing plant poison ivy. A simple mnemonic helps distinguish between the two: "Leaves of three, let it be; leaves of five, let it thrive." (Poison ivy and the related poison oak plant each have leaves in clusters of three.)

Sphinx moths like the Virginia creeper sphinx often have long abdomens that stick out past the ends of their wings. Their caterpillar forms are called hornworms. The "horns" are on their hind ends, but apparently the term "pointy-bum worm" has not yet caught on. Plants attract moths by releasing volatiles that smell good, at least to moths; even the nectar itself can train moths to want to come back by being not just sweet, but tasting like that same attractive scent. Worldwide, sphinx moths are a successful, diverse group; there are sixty species in the state of Missouri alone.

To probe deep into tube-shaped flowers, some moths have evolved record-setting tongues. According to *Guinness World Records*, the longest insect proboscis in the world belongs to Wallace's sphinx moth (*Xanthopan praedicta*), which is native to Madagascar. That moth's tongue measures more than eleven inches long—four times the length of its body. This remarkable tongue "enables it to reach the nectar deep inside the star-shaped flowers of the comet orchid."

► Eye to eye with a velvet-bean moth.

►► This is a more typical view of a velvetbean moth; changing perspective makes it look like an entirely different animal.

Not everyone ignores moths, of course. One insect-literate reviewer of this chapter wanted to share the reaction that to her, "Luna moths look like they're wearing a beautiful flowing ball gown with a big train trailing behind, and Madagascan sunset moths look like those cool samba dancers that wear bright colors and spin around." (She also wanted to remind me that there are just as many dull brown butterflies as there are dull brown moths.)

A museum-grade exception to the "moths get no love" claim can be found in the world of photography. Retired Princeton professor Emmet Gowin has had a distinguished fine art career, during which he has received multiple awards and has been featured in galleries around the world. He has enough status that he now can work with any subject matter he wants, from top fashion models to the world's most elite cultural sites. Yet modest subjects still attract him, and one special love is moths of Central and South America. He photographs them in the rainforest and then arranges the portraits into grids, focusing on color and pattern. In his celebratory book *Mariposas Nocturnas*, Gowin explains how he came to make a twenty-year project photographing tropical insects. "Over time, I have begun to see the moth as a living wonder," he says in the introduction. The moths are "visually stunning, endlessly varied, mysterious, sometimes useful, sometimes destructive, hard-working, biologically inventive, and nothing less than a miracle."

What Else is Out at Night?

Other insects besides moths carry out after-dark pollination. Most sweat bees in the tropics are diurnal, but *Megalopta genalis* from Panama—it has no common name yet—is nocturnal, and in fact has eyes that are twenty-seven times more sensitive to light than the eyes of day-flying bees. This bee collects pollen from many tropical plants, including acacias, hog plums, pochote, and the kapok or silk floss, whose seedpods contain long, silky fibers that were used to fill life jackets in World War II.

Insects are everywhere, and to find out what's flying in your neighborhood, there's a simple experiment you can do. An ultraviolet light bulb and a white sheet can be used to carry out your first nocturnal survey. Put out your light after sunset at the buggiest time of year—in northern latitudes, this will be midsummer usually, though in more tropical places like Florida or Costa Rica, any time of year is fine—and set up your light and sheet as close as you can to a meadow, forest, or marsh. Live in the city? Try it anyway: You will be surprised by how much is around.

An attentive person is all ears, but this nocturnal bee from Panama has a face that is all eyes.

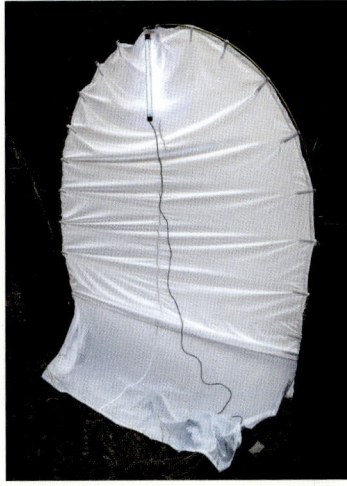

▲ An ultraviolet "bug light" will help you survey nocturnal insects.

◀ This jewel scarab is one of a large complex of related beetles that are drawn to "bug lights" in the Southwest.

In all locations, a moonless night will be more productive than one with a full moon, since no matter how good your light is, you will have trouble competing with the giant porch light in the sky. In town or out in the country, with a moon or without, do be sure to latch the screen door if you have the bug light set up near the house, since whatever you attract, you probably want to keep it outside, not inside.

There are lots of smartphone apps you can use to identify what you're seeing. One that many people find useful is Seek by iNaturalist. As it searches the databanks for possible matches to a photo of your insect in question, it will encourage you to create an account and post your sightings, geotagging them with the phone or manually entering the location using a zoomable map. If you have observed something you're not sure about, post it anyway and the community of reviewers may have some ideas. It's okay to start broad—label your image as "insect" or "beetle"— and let others help make finer distinctions. Range helps, too; you do need to know where you are, at least approximately.

Anybody can create an account and start posting; spending time benefitting from the power of our collective knowledge (and our collective goodwill) helps many of us feel optimistic.

It's easy to focus only on bad news, but the truth is, most people are smart, helpful, and surprisingly well-informed—at least about moths and beetles.

People do watch moths (the way birders watch birds), and some moth trackers use traps to see what has visited overnight. A British bird and moth fancier named Sacha Barbato talked to me about his own moth study. He told me that "the convolvulus hawk moth is a good indicator of Autumn migration in the UK. It feeds on *Nicotiana* flowers (among other plants) and is always an exciting find when seen on walls near the coast. Looking for this one makes me pay closer attention to the weather, since this moth is usually associated with winds from the Continent." Some of the moths he finds near his house in Norfolk, England, don't have common names because they are so small and poorly known. "I am always glad to see them, even so. A moth doesn't need to be huge and flashy to have an interesting story to tell."

How Big Can Moths Get?

Will there ever be a moth as big as a bus? Horror movies aside, the insect body plan comes with inherent limitations. The weight of the exoskeleton (and the lack of waterproof membranes at the joints) means gravity and fluid loss combine to limit growth. Lungs also restrict insect growth—they restrict it in a major way, since an insect doesn't have any. Lacking formal lungs, insects depend on the diffusion of oxygen across the tracheoles, which are ultra-fine tubes that connect cells. Larger insects can flex body walls to help circulate air through the tracheal system, but that soon reaches a maximum limit and stops being effective. Meanwhile, with larger size comes increased visibility; most raptors would not bother to differentiate between a mammalian squirrel and a moth the size of a squirrel—food is food is food.

We have been trained to think of insects as creepy, even monstrous. In Kafka's "The Metamorphosis," Gregor Samsa famously wakes up one morning and is a huge beetle—Oh, the horror, *et cetera*. We like the supposed "industry" of ants, or we do until they grow so bold as to investigate the kitchen, and then we are quick to spray a douse of poison on the offending foragers. Because of this, instances of parental care among insects strike us as odd, maybe even fictional . . . "Did you think I am so gullible as to fall for that?"

Yet these instances occur, as seen in the above photo of an adult spider with her babies. According to Dana Wilde, spiders caring for their young is not uncommon. He says that "some spiders even feed their young by capturing bugs and bringing them to the babies, or in some cases regurgitating food and actively feeding them. Some spider species lay a special supply of eggs, called trophic eggs, for the spiderlings to eat. A species of crab spider in Australia, *Diaea ergandros*, provides her offspring with captured prey, and studies indicate that, apparently uniquely among spiders, she recognizes her own children."

White Sands National Park: A Case Study

The ground in White Sands National Park is covered with gypsum, not snow; here, a soaptree yucca waits for moths.

White Sands National Park in New Mexico preserves the world's largest gypsum dunefield and helps educate the public about the value of dryland ecosystems. Nearby is the city of Alamogordo, famous for the development of the atom bomb, White Sands Missile Range, and a population of African oryx, an antelope native to the Kalahari Desert that was brought to New Mexico to provide exotic trophies for local hunters. That means this park preserves a unique ecosystem yet also offers a window into the Cold War, the space race, and America's complicated relationship with hunting, all contained within a spectacular panorama of white dunes and flowering yuccas.

This park is "moth city." So far, six hundred species have been identified there, and of those, 10 percent were new to science as of the past ten years. In terms of a single location, White Sands National Park has the highest percentage of endemic moths of any place in North America. Other places do have high moth diversity, like the Pine Barrens in New Jersey. As moth expert Eric Metzler points out, "That's a system, an entire ecological system. White Sands is unique in the fact that it's a single location." Metzler is discovering on average six new species a year, and that's just in an area of a few square miles. The six hundred total does not include all of the ultra-micro moths, smaller than one-sixteenth of an inch. Those still await cataloguing and study.

Not all moths here are white, and Metzler wonders whether those that are dark gray don't get a boost by warming up in the winter sun. Winter days are short; he hypothesizes that darker-colored moths warm up as quickly as they can during the day in order to fly all night. Desert winters can be cold. He observes that these moths "fly even when the temperature reaches below freezing, so somehow, solar heating or not, they have obtained enough energy to do that."

We know that Joshua trees need moths to be pollinated. The yuccas of White Sands do, too. Further research is needed, but it seems that for each species of yucca, even if there are a number of look-alike moths, in the end, only one species does the pollinating. Metzler comments: "Moths are key pollinators here.

In fall, a soap-tree yucca's seedpods crack open, providing food for rodents.

The yucca plants would go extinct almost imme-diately if it weren't for the yucca moths. There are night-flying flies as well. There are transparent bees that fly at night. There are not very many of these. We tend to think only of butterflies or bees or flies as pollinators, and of course they are critical to pol-lination, but the moths are doing the main work."

The Apache pocket mouse eats the seeds from the yucca that was pollinated by moths—thus works the circle of life.

Plants provide food and shelter for other components of the White Sands ecosystem. Hawks, snakes, and owls need mice, for example, but mice need seeds, and the seeds first need fertile plants—plants that were pollinated by moths.

The Apache pocket mouse shown above is a regional spe-cialty. This small and easily overlooked rodent, like others in its genus, gets the name "pocket mouse" from storage pouches in its cheeks. Pocket mice spend the heat of the desert day in their cool burrows, coming out at night to forage for seeds, insects, and leaves. They do not need to find a stream to drink from, because they extract water from the seeds they eat. Most pocket mice are brown, but the Apache pocket mouse has evolved to be much paler than usual, to blend in with the gypsum "sand" found inside the national park. Reptiles can be white-adapted too, so if you're in the park during the day, keep an eye out for a bleached race of earless lizard, the little white whiptail, and the paler-than-typical southwestern fence lizards.

Raider Bats and Nectar Guzzlers

Some night feeders are larger than the largest moths. Nectar-seeking bats can make very successful pollinators; saguaro cacti, for example, are bat-pollinated. This is more common than most people know: Fifty-plus bat species primar-ily eat nectar and pollen. (Hummingbirds can't eat pollen; their systems can't digest it. They supplement nectar with insects.

But some bats can and do eat pollen, and their faces and tongues are shaped so they can lick it right off their fur.)

To attract bats, plants follow the same recipe as when they want to attract moths: hosting large, pale flowers; opening at night; fragrant smells, including smelling like rotting fruit; and providing copious nectar. More than five hundred species of flowers worldwide, in at least sixty-seven plant families, rely on bats as their pollinators. Historically, that included the agaves that mescal and tequila come from, although modern plantations use cloning instead of natural reproduction to propagate their cultivars. In fact, commercial agaves are harvested just when they are about to bloom but before they send up their stalks, in order to concentrate the sugars in the plant's core. To make tequila, the heart of the blue agave is steamed and minced before being fermented. This pits short-term profit against long-term plant health: It would be better for the industry to allow some plants to mature all the way to efflorescence, as this would guarantee genetic diversity while simultaneously sustaining a base population of bats. One can say about bats what is said about insects, too: "You will miss them when they're gone."

Further, it's not anteaters but bats that are the mammal with the longest tongue-to-body ratio. The winner here is the tube-lipped nectar bat, whose tongue is 150 percent of the length of its body. When not needed, the tongue retracts in a coil, like a carpenter's measure going back into its case. The long snout of a nectar-feeding bat slots into the flower's deep well like a key

An Underwood's long-tongued bat visits a wild banana in Costa Rica.

fitting into a complicated lock, ensuring that the bat's furry head becomes fully covered in pollen.

How do you find flowers in the dark? Bats can see—contrary to urban legend, bats are not blind—and of course they also can smell their way through the forest. In the dance between the pollinating and the pollinated, echolocation matters, too. Some plant species have evolved acoustic features in their flowers that make the echo of the bats' ultrasonic call more conspicuous to their potential pollinators. These flowers often have a bell-shaped, concave form that effectively reflects the sounds the bats emit, enabling the bats to find flowers amid the dense growth of tropical rainforests. Other plant species have evolved flowers that hang in long stalks away from the plant's stem, also making it easier for bats to distinguish the flowers via echolocation. Bats are smart and adaptable, and in places such as Madera Canyon in Arizona or the Asa Wright Nature Centre in Trinidad, nectar-slurping bats have learned that the best kind of loading is freeloading. Nature lodges at these sites put out multiple hummingbird feeders. Once the sun sets and the birds go to roost, the bats show up. A bat will dive in and brake to a shuddering halt, jab a bristle-coated tongue into the hole, and peel off with a big gulp of nectar—all in less than a second. Bats' serrated tongues soak up nectar like a paper towel wrapped around a dowel. This

Lesser long-nosed bats are common "feeder raiders" in southeastern Arizona. This shot shows two bats hitting a feeder at the same time.

feeder-poaching makes at least some mammal watchers happy. It's not that the mammal fans have any animosity toward diurnal hummingbirds, but in areas where there are multiple bat species possible, bats that come to feeders offer photo ops that help clinch identification. During ornithology's shotgun era, the guiding principle was "what's hit is history, and what's missed is mystery." That motto still applies to photographs of look-alike bats.

What's Next?

Did you see any of the recent news reports about the tree frog that visits rainforest flowers? Researchers have just confirmed that Izecksohn's Brazilian tree frog pollinates the flowers of the milk fruit tree. More study is needed, but circumstantial evidence for this is strong. Passive, "on all the time" camera monitoring stations could help in verifying this thesis, as they have in South Africa for a different animal. Lizard pollination there has never been directly seen by humans, but it has been caught on camera.

Cameras may be the next frontier in night-pollination study. My friend José and I recently had camera traps set up in a nature reserve in Southern California, in an area with six resident mountain lions. We hiked in far off the trail and carefully arrayed our gear. We were going to leave it on "autopilot" for two months. What kinds of nature would visit this canyon during that time? We hoped for great puma shots.

We didn't get the puma shots, but we did get photos of other, less expected things, including a bobcat trying to lick the lens, hundreds of shots of the same kangaroo rat surveying his property, the shadow of a bear blocking the lighting units, and the shooting stars of moths that had tripped the sensor and were blazing white in the dual units of our automatic flash. In the end, we got good pictures—they were just not the pictures we had been expecting.

Kit Foxes and K-Rats: Deserts around the World

While waterfalls and sunsets may dominate Instagram, some of us prefer more austere environments. In the movie *Lawrence of Arabia*, Peter O'Toole's Lawrence gives a classic answer when asked why he likes the desert: "Because it is clean," he tells Bentley, the reporter from Chicago. As somebody who lives in a desert himself, I would add that deserts are also honest. You know what the weather is going to be (hot in summer during the day; cold in winter at night), and by looking at the orientation of the arroyos and the alluvial fans, you can know which direction water will flow, if and when it rains. Everything is exposed, especially emotions and geology. Major religion comes out of the desert, or at least it does in the Eurocentric narratives—Islam, Christianity, and Judaism all began as desert religions. So did the faith systems of Australia's first people, the Ancestral Pueblo in the American Southwest, and the Nazca people (creators of the famous Nazca Lines) in Peru. (This claim ignores Hinduism, Buddhism,

Localized summer rain falls on Zion National Park.

Shinto, and Confucianism, among many others.) You need to be self-reliant to survive in the desert, and some people move to deserts to avoid the clang and clamor of humanity. As the old saw puts it, all of the plants in the desert have prickers and thorns, and so do all of the people.

Desert plants have alarming names like bitterbrush and fish-hook cactus; the word *cholla* (CHOY-uh) comes from an archaic Spanish word for "head," as in, the head of a battle mace. Cholla segments easily detach, and when they do, they can look like a medieval weapon, especially if you have one caught on your ankle or the back of your hand. Cholla spines have microscopic barbs—easily in but not easily out—and cholla stems are one reason desert hikers keep needle-nosed pliers in the car. Datura, mentioned in the pollination chapter, can be called devil's trumpet and moonflower but also locoweed, since it contains dangerous alkaloids that cause hallucination and death. All parts of the plant are toxic.

Place names advertise deserts' hostility—from them to us and us to them. New Mexico has a desert park, El Malpais, whose name comes from the Spanish words for "bad country." Death Valley may seem to be well-named since people do die there every year. Usually, though, it is speeding, not heat stroke, that is the cause of fatalities in Death Valley. Sometimes you wonder if

▼ As morning light hits a cholla, its "don't touch!" spines show up clearly.

▲ A curve-billed thrasher stares defiantly from its nest in the middle of a stand of cholla.

verbal hostility is a kind of reverse bragging. For example, Australia has two "Little Hells," one in Tasmania and one in South Australia. (It also has a Purgatory Hill and a Devil's Lair.) Devil's Punchbowl is a shale-and-sandstone park outside of Los Angeles, and you could stop there on your way to hike in the Devil's Cornfield, Death Valley. Death Valley National Park famously is the home of Badwater Basin (another grim place name), where you can stand 282 feet below sea level. The water does indeed taste very bad there. With a long lens you can photograph the "sea level" sign bolted to a cliff high above the car park.

Desert wildlife has adapted to the harsh conditions in many interesting ways, of which the most universal is to switch the wake-sleep cycle and become nocturnal. That's good for wildlife spotters, since on average, that's a nicer time for us to be out, too, plus there's usually less traffic on backcountry roads at night. If the night is a bit cool, snakes like to warm up on the pavement, while low vegetation and open vistas make using thermal scopes and spotlights especially productive. Depending on where you are, there may not even be a single gnat, fly, midge, or mosquito.

Water: Too Much, Too Little, Sometimes Just Right

In recent years, torrential summer storms have washed out roads and flooded campgrounds in Death Valley. Deserts typically do not have any duff in their soil profile—no fluffy layers of semi-decayed leaves and animal remains that create humus, recycle nitrogen, and soak up rainfall. In the desert, too much rain falling too quickly does not have time to soak into the hardened, gravelly soil, and so it sheets off, building up into churning floods of brown water, rocks, and debris as the water funnels into the nearest canyon and firehoses out the bottom, washing away anything in its path. These storms can be very localized. There can be a flash flood downstream caused by rain that the flooded area never saw, smelled, heard, or felt. (One term for a localized flash flood is *gullywhumper*.) We see that in this picture of Zion National Park, where it is raining in one part of the canyon system but not in the other. That water may become a dangerous surprise once it reaches a downstream terminus.

Localized summer rain falls on Zion National Park.

We're used to thinking about precipitation as measured in inches (or centimeters) of rainfall. Desert animals know that is too simplistic. In the Atacama Desert of Chile and Peru and the Namib Desert of Southwest Africa, sometimes the only precipitation is the accumulated droplets of coastal fog. Forty-eight kinds of animals in the Namib Desert "drink" fog, which can bead up on the thorns and stems of plants to concentrate into magical pearls of non-rain fresh water.

Other places, the water sources in the desert are human-made. In Australia or the American West, stock tanks provide water not just for cattle, as intended, but secondarily help toads, bats, dingoes (or coyotes, depending on which continent), feral burros, feral camels, swallows, quail, and bighorn sheep. Even bees and wasps can use stock tanks—something to keep in mind if you're allergic to their stings.

The usual punch list for making an artificial waterhole is simple: Drill a borehole, pipe it and stick on a valve, erect scaffolding, and attach a rudder and a ten-foot fan at the top, jointed so the unit swivels freely. The vane turns the blades of the fan

▶ This historic photo shows a windmill and stock pond; cattle use the pond during the day and wild animals at night.

▶▶ Bees at dusk swarm around a stock pond.

so they're always facing the wind, and as the windmill spins, it pumps water up and delivers it to a trough, pond, or muddy wallow. With maintenance, these structures can last for many decades. There can be variations; electric pumps (run by discrete solar panels) are used in some wilderness areas to help animals like bighorn sheep during droughts. Sometimes fencing is used to exclude domestic stock in that case, or at least to limit the amount of vegetation they trample and the amount of urine and feces they introduce into the standing water.

Desert animals are expert water seekers. I have watched midday lizards drink from a leaky hose in Arizona and midnight skunks find the last summer puddle pooled up inside a culvert under a busy roadway. In Oman I have seen bats drink from a hotel swimming pool, though on average, they seem to prefer a natural slough or even an agricultural ditch. Raccoons surprise me by thriving in my desert town, only coming out in the quiet-est hours of the night and apparently making use of such "wet" places as storm drains, golf courses, overwatered backyards, and the runoff ditches along railroad embankments. How much gen-eral foraging they are doing versus how much raiding of urban sources like unattended dog food, I am not sure. Adult raccoons weigh between ten and twenty pounds, and their nimble hands can open food wrappers and shed-door latches, so whatever they have figured out to do, I assume they are doing it quite well, since their population stays stable, year after year. As a brief sidenote,

I love the Dutch word for "raccoon," which is *wasbeer*—"wash bear." Raccoons are not native to Europe, but because of escapees from fur farms and released family pets, they are now established there, as well as in Japan, Russia, Armenia, and the Republic of Georgia.

These examples of the animal-suburban interface are shared here intentionally, as a reminder that nature is everywhere, and that contrary to the common perception, there is not the pure "out there" kind of nature at odds with the contaminated, value-less, impure "around here" kind of nature. Nature is nature, and the raccoons raiding my neighbors' rubbish bins at three in the morning do not have some kind of inferiority complex because they can't be true and noble Yosemite Park raccoons. If they could see a map and make a list of options, my local raccoons might not want to be in Yosemite or Sequoia instead of my desert, but rather living the good life in Berlin.

What about those animals who have solved the water problem without resorting to dog food and storm drains? A great example of deserts-at-night evolution can be seen in the Great Basin spadefoot. This desert animal looks like a small toad or bumpy froglet, although technically it's neither. It looks like a toad but is its own thing, a "spadefoot," named for a little spur on the hind leg that it uses to dig into the mud.

In the American West, summer rains can be hard and brief, and they create surface ponds that may be gone in a few weeks or months. Spadefoots have evolved to exploit these brief gifts of water. They enact the same life cycle as any amphibian (egg, tadpole, adult) but do it all at triple speed. Adult Great Basin spadefoots forage at night, feeding on beetles, flies, wasps, butterflies, moths, dragonflies, and spiders. Late-summer rains trigger the breeding impulse. Males call, females make their choices, and up to a thousand eggs are laid. Eggs can hatch after just a day or two, and tadpoles can be metamorphizing into adults by the two-week mark.

Once hatched, tadpoles can either be omnivorous, grazing algae or taking what detritus and fairy shrimp the pond randomly allows, or they can be carnivorous, with a larger head and more advanced beak. After a brief life in water, they become adults, they eat and eat, and then as early as a month later, they go underground to wait for the next set of storms, even if they have to wait a full year. The burrows they dig are one to three feet down and backfilled with soil.

While in torpor, spadefoots avoid drying out by absorbing moisture from the soil. Most go into solitary burrows, though there are reports of sharing, and there are also a few reports of spadefoots using rodent burrows instead of digging their own direct and private foxholes. When the rains come back, the male choruses a quacking call, males and females join in the water, and the eggs are fertilized as they leave the female. The next generation is ready to carry on.

Another fabulous night animal is the North African elephant shrew. In general, I am not into exotic pets, but if these were easily available, I would want one. The name in this case is entirely accurate. Elephant shrews are indeed related to elephants, except they are carnivorous and hamster-size and hunt like large, hyperactive shrews. Another name for this animal group is sengi (also spelled tsengi), or plural, "sengis." There could be as many

▼ The spade-foot is a small, toadlike amphibian that breeds in temporary rain pools and then burrows underground.

◄ This hamster-size curiosity is a North African elephant shrew.

as twenty species, found only in Africa in savanna habitat and more tropical forests; the North African kind, as per the name, is the rocky scrubland one and is found in Algeria, Libya, Morocco, and Tunisia. They can eat worms and fruit, but for this species, dominant foods include pavement ants, termites, and wood lice. The shrew's long nose can turn in a circle—it's not a full trunk, but it is adroit and sensitive and makes this animal very good at snuffling out dinner. This animal does not hibernate fully, but when food is scarce, it can enter a suspended state called torpor.

You may have seen this animal before and not known it. In classical Egyptian mythology, deities are portrayed with animal characteristics. Jackal-headed Anubis is a guide to the under-world, and Horus, god of the sun and sky, has a falcon's head. What to make of Set, though, the god of deserts and storms? His head looks like an anteater's, but that doesn't make sense, since the giant anteater only inhabits South America. Despite the dura-bility of papyrus rafts, demonstrated by Thor Heyerdahl's 1970 Ra expedition, we hold the position that the ancient Egyptians had no knowledge of jaguars, sloths, llamas, or anteaters.

A case has been made that Set's head design is based on the elephant shrew's profile. Alternately, Set (also spelled "Seth") could be an aardvark, which is found as close to Egypt as present-day Sudan. Or it could be a sort of droopy jackal, or a composite of multiple animals, or not an animal at all, but an act of creative imagination. Adventure movies like to bring mummies back to life, but once these mummies are revived, the modern characters never stop to ask the important questions, such as, "So what's with that god Set, anyway?"

Taxonomy is the study of how animals are related to each other on the tree of life; its aim is to figure out which species most closely belong to which groups. The North African ele-phant shrew was recently separated from other elephant shrews and placed in its own genus. This argument was based primarily on a study of DNA. One final detail that helped make the case?

Not to be indelicate, but the shape of the boy shrew's gentlemanly bits—once they were examined closely—turned out to be very different from those of the other species it was previously grouped with. When it comes to science, even the smallest detail matters.

The kangaroo rat is a champion at doing without water. Or rather, "rats" plural, since there are twenty species. The "k-rats," as biologists affectionately call them, are neither rats nor kangaroos. These rodents live in the deserts and borderlands of western North America, and they all look generally the same. The generic k-rat has a round head; a short, tan body with white stripes; spring-loaded back legs; and a *looong* tail with a fluffy banner at its very end. The tail is always longer than the head-plus-body length; the three measurements together add up to a combined total of twelve inches or more. K-rats run by dashing quickly from creosote bush to creosote bush, a "run" that is an accelerating series of dash-hops, almost too fast to see. Dazzled by car lights or simply curious, a k-rat will sometimes pause along the side of a quiet road, and if an observer approaches slowly, they can get close enough to touch.

Even if dazzled by your flashlight, kangaroo rats will know you are there. Such good hearing! They can even hear owls swooping down—or certainly hear the patter of oatmeal hitting the ground, if I am being messy in camp and "accidentally" drop some food because I want to watch them come closer to my truck. Researchers who lure them into non-harming box traps for study find that k-rats are very fond of a mix of oatmeal and

peanut butter. I use that mixture for wildlife photography, too, but I usually add pepitas (shelled
pumpkin seeds).

   This animal is more athletic than it looks. A
kangaroo rat can jump nine feet in one leap—the
tail is both rudder and counterbalance—and they
can even turn around and kick sand in a rattlesnake's face. Their most special skill, though, is the
way they survive without water, getting much of
their moisture from the seeds they eat. Spending the day underground helps, of course, since burrows are not just cooler but
often have higher humidity than the outer air, further lessening
water loss. K-rats do not sweat or pant, and their ultra-efficient
kidneys concentrate urine to a viscous paste. Claims online say
they can live ten years and not drink water once; these same
sources also say they sometimes eat insects. I've never seen that
myself, but it sounds plausible. I have, however, seen them in midsummer in the heart of Death Valley, in habitat with no aboveground springs—so they certainly are getting by just on seeds
and other plant matter most of the time.

A kangaroo rat poses inside the author's portable studio in the Mojave Desert.

   Kangaroo rats are nocturnal (hence their large eyes), and
they also can see a range of light that humans cannot, since their
vision extends into the ultraviolet spectrum. Ah, but why? Possible answers include (a) this extra capability is something they
inherited from an ancient gopher ancestor about twenty-five
million years ago. Or (b), since rodents mark territory with urine,
UV vision helps them see and not just smell these social markers.
Or (c), UV vision gives them an advantage during crepuscular
foraging, since the relative proportion of UV to visible light is
higher at dusk and dawn than it is at midday. Or (d), it helps
them see brown seeds in brown dirt, since under UV light, seeds
can look different from the soil substrate. Or perhaps (e), it helps
the k-rats distinguish up from down. Bats can tell "high in the
sky" UV light versus "closer to ground" UV light, to help them

maintain an imaginary horizon line; maybe with k-rats, there is something like that, too, for example if they look up and need to distance-gauge an approaching owl.

But wait, there's more—option (f): It has some function in navigating the dim light of burrows and/or navigating out in the open on moonless nights.

Or (g), all of the above.

Or (h), none of these choices and we have no clue what's going on.

That last option explains (in that it doesn't explain) North American flying squirrels, which comprises a cluster of three related nocturnal species (genus *Glaucomys*), all of which fluoresce in UV light. Why do these squirrels glow in the dark? There are lots of possibilities and no firm answers, which may as well be the subtitle of this whole book.

Temperature (But It's a Dry Heat)

Being water-wise also ties in with being nocturnal, which also ties in with the ways that desert animals deal with extreme heat. Thinking about these adaptations also allows us to consider the ways that convergent evolution has turned unrelated animals (often living on continents far away from each other) into "brothers from different mothers."

This desert spiny lizard will be more active once it has had its "coffee"—a good dose of morning sunlight.

The conversation about heat starts with the acknowledgment that for some creatures, heat is something to be courted, not avoided. Cold-blooded lizards rest at night and come out by day, basking first thing to let the sun help jumpstart their metabolism.

More typical is the jackrabbit, which is a desert hare from a widespread, worldwide group. Aesop's "The Tortoise and the Hare" involves a jackrabbit relative; the common names for other members of the genus *Lepus* show their diverse habitats and ranges. Among others, there are Ethiopian hares, Korean hares, Burmese hares, Chinese hares, and, in Mongolia, Tolai hares. Jackrabbits are just hares with desert adaptations, most visibly their extra-long ears. The Tehuantepec jackrabbit is endemic to Oaxaca, Mexico, and has a lower half that is all white, as if it hopped into a vat of white paint and jumped out again without going the rest of the way under.

The black-tailed jackrabbit is the most commonly seen hare in North America. Expect it in open, dry habitat—deserts, scrublands, sand dunes, prairies, and overgrazed rangeland. This hare's tall ears do two things. Enhanced hearing is one, but more important, tall ears allow the hare to regulate its body temperature because they are full of blood vessels. When the jackrabbit gets too hot, these blood vessels widen, which dissipates excess

➤ Tall ears help jackrabbits radiate excess heat.

▼ Jerboas combine long legs and tall ears in the same package, but they are not closely related to jackrabbits or kangaroo rats. This one is from Mongolia.

heat. Having eyes on the sides (not front) of their heads gives them poor binocular vision but a wide field of view on each side. They may freeze if they sense danger, trying to blend in, but when "fight or flight" becomes "run away," their long back legs accelerate quickly into bounding zigzags.

Convergent evolution means that different lineages of animals end up "inventing" the same solutions to environmental conditions. Tall ears and springy back legs can be seen in other, non-hare, non-jackrabbit kinds of animals. Jerboas, for example, are found in North Africa and Asia and look like a mad scientist's mashup of a kangaroo rat and a jackrabbit, resulting in an animal with long ears, a long tail, and large, coiled-spring hind legs. Some people think they look like tiny kangaroos with Easter Bunny ears tied on. If you have been on an African safari and done a night drive, you may have seen another bouncy rodent that jerboas resemble: the springhare. Even though many people in North America are not familiar with jerboas, worldwide there are over thirty species, ranging from the mouse-size Baluchistan pygmy jerboa, endemic to Pakistan, to the seventeen-inch-long great jerboa, which is native to Russia, Ukraine, and China.

The jerboa (or "desert rat") was the symbol of a famous British tank unit in World War II, the 7th Armoured Division, which was directed by Field Marshal Montgomery against the German general Rommel, who was known as the Desert Fox.

In the end, the rat beat the fox.

Our final example of desert jumpers comes from Madagascar, an island famous for its multiplicity of habitats, including xeric forests and tropical thorn scrub. The island's largest rodent is large indeed: The Malagasy giant jumping rat is two feet long, counting its tail. It comes out at night, resting during the day in a tunnel system than can be sixteen feet long and have as many as six entrances. These jumping rats are popular in zoos, which are trying to create breeding centers, but in the wild, enthusiasts go to one place to see them: Kirindy reserve, a forest in Madagascar and where this picture was taken. We will return to Madagascar in a later chapter when it is time to think about nocturnal primates.

While long ears are good to have in the desert, they would be bad in the arctic, where they could too easily be damaged by frostbite. In 1877, naturalist Joel Asaph Allen published a paper stating that animals that live in colder temperatures have shorter and thicker limbs than animals in warmer environments. Thicker features change the volume-to-surface ratio, allowing cold-adapted species to retain heat and the warm-adapted ones to shed it.

Allen postulated this after observing the lengths of arctic hares' limbs and ears and comparing these measurements to those of other North American rabbits in the same genus. Allen's rule applies to other animal groups besides jackrabbits and is often illustrated by pointing out that arctic species are stockier than their tropical counterparts (e.g., a wolf versus a coyote, or a moose versus a pronghorn antelope). A husky is built for the snow; a whippet is not.

Foxes—found from the high arctic to the hottest deserts— enact Allen's rule very clearly. The arctic fox has a short muzzle; blunt, rounded ears; and a thick tail. It can curl up in a ball to sleep, covering its face with the fluffy tail. The kit fox, native to North American deserts, needs to leak heat, not conserve it, and so it is small, slim, and much longer-eared than the arctic fox. It has a longer muzzle and a pale tan coat.

Other species of wild canid share the kit fox's desert landscape, which has influenced size as well. Originally it had to compete with "lobos" (the Mexican wolf, now mostly extirpated) as well as coyotes. It survived by being small; kit foxes can eat things as small as crickets, and they also eat cactus fruit, roadkill, and kangaroo rats. Like k-rats, kit foxes can survive without free water, gaining what they need from the blood and moisture in their prey. Kit foxes are nocturnal, especially in summer, and retreat into the coolness and humidity of their dens during midday.

The Malagasy giant jumping rat continues our trend of "long-eared jumping things." It is endemic to Madagascar.

With ears as big as its face, a young kit fox emerges from a den under an abandoned house.

Their name conflates several strands of language. "Kit" means baby animal or cub, and for this diminutive fox, it also may be a shortened form of kitten, since even an adult kit fox is still only the size of a robust house cat.

Spiders and Scorpions (and One Gecko)

The largest spider in the world is in Peru . . . and it is made of rocks. It is part of the complex of geoglyphs called the Nazca Lines. These are human-made spirals, designs, and miles-long lines in the Atacama Desert. The figures are two thousand years old and cover an area of 200 square miles, and I will say, having been there, that they do need to be seen from the air to be fully appreciated. Besides the geometric shapes, there are also animals—a hummingbird, a heron, a spider, a monkey, and so on. If you have a vivid imagination, there is also an astronaut. Depending on the sources you turn to, the lines may have been constructed as irrigation schemes, fertility rites, star charts, or directional aids for visiting spaceships. Some are obscured now by erosion and motorcycle tracks; with enhanced aerial surveys, new lines are identified every year.

The oldest desert animal, at least in terms of continuity of form, is the scorpion. There are at least 1500 species in 170 genera, all of them filling out a body plan that first emerged 420 million years ago and has stayed more or less constant right up through today. The scorpion's pincers are called pedipalps by entomologists (and Scrabble players); like spiders, scorpions have eight legs, and of course scorpions come with the famous stinger cocked and ready at the end of the tail (which is called the "metasoma"). This is both an offensive and defensive weapon. A scorpion uses its stinger to zap a mouse or insect, subduing it, and then may strike further to take prey from "stunned" to "all the way dead." Usually, a scorpion prefers to use its pincers for initial capture, since venom is costly to produce. Enzymes in the mouth help it liquify prey once something has been caught, because a scorpion has no teeth and can't chew. It never gobbles dinner, but instead slurps and sucks.

In North Africa, the deathstalker scorpion is supposedly the deadliest of all, and even if not, it certainly has the coolest name. With the forty North American species of scorpion, the sting is painful to most people but (on average) is not deadly—I have been told it is no worse than a bee sting, but I'd rather not find out firsthand. Usually, the scorpion's second strike has less venom than the first, so if you're going to get stung, wait until a friend gets zapped first. That said, two of the forty North American species have more potential to be deadly than others, and with all scorpions, allergic reactions are possible. Prudence is best: Look but don't touch.

The best thing about scorpions is how easily they can be seen at night with an ultraviolet flashlight. A fluorescent material found in their outer shell makes them glow in the dark, like the giant hairy scorpion shown on the next page. Why do they glow? The short answer—and this is a direct quote from *National Geographic*—is that "no one knows." But of course, not knowing has never stopped curiosity before. Many theories have been

posed to explain why scorpions glow: as a result of random, accidental chemical reactions; to lure prey (though the opposite effect is more likely); to warn predators; and/or to "convert the dim UV light from the moon and the stars into the color that they see best—blue-green." The idea is that they use that data to find nocturnal shadows and hide from predators, such as the scorpion-hunting pallid bat that we will meet on page 207.

National Geographic expands on this last hypothesis, saying that a researcher named Douglas Gaffin "thinks that the scorpion's entire body, from the stinging tail to the crushing pincers, collects UV light from the environment and convert[s] it into blue-green wavelengths. These signals could even pass to the brain via clusters of nerves that are spread throughout the animal's body. If this idea pans out, it means that a scorpion's glow could increase the surface area of its eyes by a thousand times. The entire scorpion would effectively be one big eye."

In North America, scorpions give birth to live young through the summer months, having retained sperm from mating the previous year. The babies are not fully developed when they are born and take one to three weeks to mature. As soon as they are born, they crawl up their mother's legs to her back, where they will ride until they molt. This habit was well-known in the ancient world. In Egyptian mythology, the goddess Isis was linked with scorpions, as she was a symbol of a devoted mother. Modern Mother's Day cards display lots of flowers and butterflies, but not nearly enough scorpions.

Some seeming scorpions are not scorpions at all, but rather are lizards. The western banded gecko is a two- or three-inch nocturnal lizard with a detachable two- to three-inch tail. According to the Arizona-Sonora Desert Museum, "Active principally at night, western banded geckos can be seen crossing roads during summer. It has been suggested that their gait and carriage mimics that of the scorpions of the genus *Hadrurus* that share the same habitat." *Hadrurus arizonensis* is the Latin name of the large, glowing scorpion on the previous page; there are half a dozen species in that genus. Is the small, nonvenomous lizard shown below trying to avoid predation by resembling a large, fierce, well-armored scorpion? If so, that makes sense, since the gecko's detachable tail might be called a weapon of last resort—it represents a huge investment of resources, and it is where the gecko stores food and water for lean times, including winter dormancy. The loss of the tail can put the survival of the lizard in jeopardy (one reason not to pick this animal up if you see one in the wild). If it is true that these lizards walk like

Active at night, western banded geckos hunt beetles, spiders, grasshoppers, and sowbugs.

scorpions for protection, it isn't the only debt they owe to their *Hadrurus* neighbors: Baby scorpions are among the many things that western banded geckos eat.

Like other lizards, these geckos need very little water, because they excrete their metabolic wastes as a solid (uric acid) instead of dissolved in water (urine).

One final insect that can attract our attention is the globe mallow bee. In the US Southwest, globemallow is a desert plant with sage-green leaves and delicate, apricot-orange flowers. (The bee's name is spelled as separate words, globe and mallow, while the plant is spelled all as one.) Desert globemallow grows in clumps two or three feet tall and about the same size around; the flowers look like small poppies. These plants are primarily pollinated by bees, though they are visited as well by butterflies (including hairstreaks, skippers, and painted ladies). Desert globemallows need little water and make a great addition to dryland gardens. The word *mallow* refers to an Old World botanical group, of which the marsh mallow plant eventually became the squishy white candy we call "marshmallows" today.

The globemallow of the American deserts closes shop after dark, and that's what globe mallow bees know to exploit. At day's end, the bee seeks out a globemallow flower, and as the blossom closes for the night, *voilà*—it becomes the bee's house, bed, and covers, all provided in one snug bundle. Assuming the insect has been (pardon me) as busy as a bee during the day, the flower receives intimate contact with an insect that is thickly covered in pollen.

Deserts and You

We've tried not to remind you on every page, but going hiking, day or night, is easy and fun, and it never has to be a long slog or a high-tech expedition. A simple loop trail around the nature center or even just staying on a bench by the car park as the other hikers pack up and leave and night arrives inch by darkening inch—these things can be satisfying and restorative. Turn your phone off and just let the sun go down and the cool air fill your

Mammal spotters in Morocco head out at sunset to look for animals.

heart, your lungs, and the back corners of your brain. What can you smell? Are there nighthawks coming out, or have the coyotes started to yippee-ki-yay? In deserts, even in summer, nights are the cool times, both temperature-wise and in the "oh my gosh, this is so cool" way.

I have been luckier than some in my nature-at-night trips, so yes, over the years I have seen a puma and a wolverine and a Pel's fishing-owl, an African species so fierce that one once ate a baby Nile crocodile. I have seen jaguars and tree kangaroos and hedgehogs and the fer-de-lance snake. In all that travel, I've had no snakebites (touch wood), and to be honest, snakes are so uncommon that there have been nights when I have been out driving and couldn't find even a small rattlesnake for love or money. (This happens most often when I have a photo deadline or an out-of-town guest.)

On paper, the coral snake is about the deadliest snake in the world, with the lethality of its venom, gram for gram, ranking right up there with the African black mamba or the Australian inland taipan. Yet when I took the coral snake photo on the next page, the snake featured in it was not any true threat to me or my photo buddy. The venom is toxic, true, but you have to mistreat this species to get it to bite you. People are wise to be cautious around any animal they don't know, but the risk of snakes in North America has been exaggerated by movies, which imply that every bush has a coiled sidewinder under it and every bite is fatal. If you want to be afraid of something, be wary of cars (forty-three thousand fatalities in the United States per year), tornadoes (eighty deaths per year), lightning (twenty deaths), hippos (in Sub-Saharan Africa, an estimated five hundred deaths a year), bison (a few selfie-taking people in Yellowstone National Park each and every year), or handguns (even more dangerous than cars, with an annual fatality number so large it almost seems impossible to believe). A few snakes on the afternoon trail are the least of our collective troubles.

Is it better to die for truth or for beauty? Emily Dickinson asks us that question in a famous poem. How about option C, not dying for either? For me, beauty is a wondrous and dangerous drug. The more I get of it, the more I want next time. So, here's a plant to wrap up our desert discussion with: *Peniocereus greggii*, the queen of the night cactus. It is a southwestern species, and most of the year it is not much more than a scraggly assemblage of twiggy branches that look like dead, gray sticks. It grows under and around creosote bushes, ironwoods, and other desert shrubs. Pollinated by hawk moths, the queen of the night cactus blooms only at night, and only for one night out of the year.

That final night, after all that waiting, it sends out a terrific-looking flower, and the flower's rarity makes seeing one all the more intoxicating. It almost seems like some kind of fantasy plant, the kind that would be in a story where the Dungeons & Dragons team must overcome obstacle after obstacle to reach the enchanted plant at just the right moment when it opens up and glows white in the moonlight. We think the evil sorceress

▼ This coral snake could be deadly if you encouraged it to bite you. If you leave it alone, there is no risk.

▲ The queen of the night cactus only blooms one night of the year.

has defeated them when, *at last*, just in the final moment, they arrive on the mountaintop and watch as the flower unfolds. Should nature come with a soundtrack? This flower should.

Yet the story of nature has many sides. Fans of truth still have work to do as well. Another constant theme in this book is how many basic facts remain to be discovered. Where does such-and-such a thing live, how does it reproduce, what are its enemies or its special requirements or what does it do during drought, fire, famine, or flood? Birds are better known than most vertebrates, maybe because they can be brightly colored and because so many come out in the convenience of daytime. Yet night birds are much less well-known than day birds, and the same goes for night mammals. There is room in the pages of science books for anybody willing to go and look, to go and measure, to go and do some very basic question asking and answering. Community science matters, and it certainly can use more volunteers. Want to make a difference? Here's your chance. The photo above shows an African rodent, the gerbil, which is somewhat related to the pocket mice of North America. But *which* gerbil is it, that's the question being asked by the researcher in this photograph—which gerbil and what was it doing hundreds of miles past where it's been found according to any previous records? Has it been expanding its range or had it just been overlooked by other surveys?

These questions and many others are still out there, waiting for a curious observer to come and explore desert nature after dark.

A researcher examines a gerbil in Morocco, documenting a range expansion for this species. It was not known to occur at this site.

Tapirs, Tenrecs, and Tarantulas: Life on the Forest Floor

All hail the snufflers and the rooters, the cleaner-uppers, the ones who glean and nibble and paw and reveal, the understory animals who live in the shadows and who take all that has fallen from above and make it useful once again.

We are glad to have an animal like the agouti, because the rainforest floor would be a rotting, stinking mess without it. Take the durian fruit for example. It's a love-it-or-hate-it kind of food. Native to Asia, the fruits can weigh six pounds and be as big around as an American football. They smell like . . . well, that differs depending on whom one asks. Durian either smells like a mix of gym socks, rancid meat, and dirty diapers, or else it is a kind of vanilla-custard-rainforest sort of experience, like an exotic apple-pear or a musky but pleasant botanical garden. I like it myself but can't afford it very often. When it is sold in Los Angeles, durian is air-freighted in from Thailand, and prices start at thirty dollars a pound. Both the love and hate sides agree that you wouldn't

An ant's-eye view of the Costa Rican rainforest.

want to spend long walking on a forest floor that is ankle-deep in rotting durian.

To understand what's going on with the forest floor, it helps to back up and think about the tropical forest in cross section. The rainforest is not just one big mass of unified vegetation. The tallest trees are usually about 150 feet tall, and their crowns form a fairly continuous canopy or "roof." That is because the most dominant, canopy-creating trees are trying to get their leaves up higher than everybody else's leaves in order to soak up all that abundant equatorial sunlight. The end of one tree's radius touches up against the leaves of the next. Yet there are problems with a "race to the top." The physics of moving water up and down inside a tall trunk limits ultimate growth, as does how much weight the roots and trunk can support. (The mosses and water-filled epiphytes that grow on tree trunks are heavy.) Any tree must balance the benefit of access to sunlight against the risk of being hit by lightning or having hurricane winds rip it out of the ground or snap its trunk in half. Most trees want to be tall, but only *so* tall. Beyond that point, being taller than your neighbors becomes a liability.

Even tall trees start out small. At the same time, other trees specialize in thriving in low light and can successfully occupy sub-canopy strata; they will never break through into the sunlight hundreds of feet up, because their leaves are not optimized for that, nor are their trunks and branches sturdy enough to sustain that much stress and wind shear. It's like architecture: Some shapes are good for igloos and garden huts, and some designs are better for cathedrals and skyscrapers. In the mid-layers there are vines, too, as well as smaller bushes and parasitic trees that grow out of other trees. The air circulates differently below the canopy, and light levels are dimmer, so leaves need to be shaped differently. The result is a distinct set of layers. At their most basic, we can divide the layers into four communities: forest floor, understory, main canopy, and emergent trees (the ones even taller than the rest of the canopy).

You won't see these layers from a boat along the river, nor even from most roads through the tropical forest. Opening up a light gap in primary forest changes the species mix. In those cases, such as alongside a two-lane highway, the "wall" of the second-growth forest can indeed seem like an impenetrable mass of green. Once one is through that immediate edge, the profile reverts to the original scheme and the rainforest once again divides into layers. Looking up, you can identify the plants that mostly occur in the low-low strata, those that are medium-low, the medium to medium-tall layer, and so on up to the general canopy. If you're on a jungle trail that includes a viewing tower, going up a hundred feet or more not only takes you into the canopy, but lets you climb above the canopy to be eye level with the final emergent giants towering above everybody else.

Different forest layers provide habitats for different animal communities, too. Native to Panama and Colombia, the Geoffroy's tamarin is a small monkey with a twelve-inch tail. Their family groups leap among the tops of trees, hunting for insects and ripe fruit. They also eat leaves and tree sap. Some of their leaps from tree to tree are quite spectacular, and they can land in saplings with a leaf-shaking crash. The double-toothed kite is a small forest hawk that hopes to catch the lizards and grasshoppers flushed out by the tamarins' ruckus, so it follows them closely. The hawk can't eat the monkeys—it is too small for that— but it can make agile turns flying under and through the canopy, ready to catch whatever gets driven into view.

In North America, we're used to thinking of hawks as large, broad-winged birds soaring over open grasslands. Panama, in contrast, has fifty-six types of falcons, hawks, kites, and eagles. One kind eats bats (the bat hawk) and another snails (the snail kite), and although the pearl kite does not eat pearls, it *is* pretty

as a pearl and smaller than a kitten. Each of these raptors is built to find its own niche. Some do fly over open country, but others specialize in under-the-canopy hunting, like the double-toothed kite shown above. Its name refers to a ridge on the mandible; like other birds, it does not have any actual teeth.

▼ This double-toothed kite studies a troop of monkeys, hoping they will shake loose a lizard or grasshopper.

▲ A Geoffroy's tamarin scolds the camera. "This is *my* branch!"

Life on the Forest Floor: Typical Animals Doing Typical Things

Gravity is a very useful thing, since without it our coffee wouldn't stay in our cups, and those of us who wear dresses would always be trying to tug our skirts back down. For rainforest animals, it also means that all the good things that are up high in the canopy, such as fruit or seedpods (and the occasional ill bird), sooner or later end up on the ground. Let the feast begin!

The fact that wads of ripe fruit and giant seedpods magically fall from the sky is good news for the Central American agouti (*ahh-GOO-tee*). This reddish-tan rainforest rodent eats things like palm nuts, often sitting on its hind legs to manipulate the treat with its front paws, looking like a cross between a rabbit and a busy squirrel. It has such strong teeth that it can crack a Brazil nut without needing a hammer or an acetylene torch. Agoutis are terrier-size and large enough to spot without binoculars. They are often out in daytime (especially early in the morning), so tourists see them more often than they do other jungle wildlife. Their gait is hard to describe; some people find them a bit piggy or scampery, though in truth they are just "agouti-like." Agoutis' alarm call is a short, sharp bark. They cache seeds for later, only in some cases "later" never comes. In the intervening weeks or months, the cache-placing agouti might have been snatched by an ocelot or poached by a village hunter, or a flood may have come and washed out the trail, or maybe that particular agouti just got a bit addlepated and forgot where it hid the stash. No matter what the cause of the owner's disappearance, this scatter-hoarding behavior helps regenerate the forest, since a buried seed is halfway to becoming a germinated seed, and a geminated seed is well on its way to becoming a new tree.

Agoutis can be seen at night (or more commonly, at dusk and dawn), but if there's enough food, they forage in the day and head to a burrow or hollow log before sundown. The evening shift is taken over by a similar animal, the paca. It is more cautious than the agouti, with rows of white dots that break up its dark shape and look like moonlight dappling tree bark. Pacas

Agoutis are tall guinea pigs that eat palm nuts, burying the extra to become tomorrow's forest.

and agoutis are similar in size and shape, but pacas weigh more and come from a different lineage. There are two paca species only (lowland and highland), compared to twelve kinds of agouti. Both groups swim well, and both animals can survive only on fruit, but the paca caches less often than the agouti and forages more broadly, also eating insects, tubers, and fungi. It often brings food back to eat near the burrow entrance, where safety is close at hand.

Pacas are monogamous. Males and females feed separately from each other and have separate burrows. There seems to be an un-rodentlike aloofness to them; in zoos they do not reproduce easily. In nature they communicate with secretions from anal glands, as well as by sound. At times, though, things become more festive. According to *Mammals of Mexico* (edited by Gerardo Ceballos), the paca's "cheeks are prominent due to the expanded zygomatic arch" (that is, the cheekbone, that comma-shaped part of a face between the bottom of the eye and the start of the jaw). This creates a space used as a resonating structure, "a unique feature among mammals." Males have larger resonating chambers than females, and the larger space "apparently amplifies emitted sounds." The University of Michigan's animal diversity website explains that "when air is pushed through the chamber, a low rumbling sound is produced." If we

Pacas are nocturnal and often hard to see, in part because they face pressure from bushmeat hunters.

could hear what the animals hear and smell what they smell, think how different the night forest would be for us.

An order of magnitude larger than the agouti is the tapir, the pig-snouted horse-cows shown at a watering hole among the apes in the opening scenes of the movie *2001: A Space Odyssey*. (Those stand-ins for prehistoric animals were live Brazilian tapirs from a British zoo.) There are four types of tapirs world-wide, three brown kinds in South America and a black-and-white one in Malaysia. Like agoutis, they eat palm nuts; like pacas, they evade predators by hiding underwater. Tapirs date back to an ancestor they share with horses and rhinoceroses, from tens of millions of years ago. Their feet resemble a hippo's. All four types of tapir are hunted and tend to avoid people, but in Costa Rica's Corcovado National Park, one might encounter the Baird's tapir, and in the Brazilian Pantanal, night drives often turn up feeding tapirs.

Some forest-floor feeders increase their foraging options by occasionally leaving the ground to investigate the lower branches of trees or by poking into garden sheds and compost piles. One attractive scrounger of this type is the white-eared opossum of South America. Like the paca, the agouti, and the tapir, it eats fruit and seeds, and while it does not bury them for later, it does disperse them through its poo. It also eats snakes, insects, crabs, and anything else it can dig up out of the wet earth; like other opossums, this species is broadly omnivorous. It also is a marsupial. When born, the dinky, jelly bean–size babies (called joeys) make their way into the mother's pouch to suckle milk and continue maturing for two more months.

Like its better-known cousin the Virginia opossum, this species also can play dead, in an exhibition of what is technically called *thanatosis*. This involuntary catatonia is more than mere feigning. The reaction creates a rigid, insensate body that gives a very realistic appearance of death. Sometimes a foul odor is released as well. The death phase may last anywhere from six

The white-eared opossum of South America may not have startlingly white ears, but depending on the light, their silver ears look whiter than the ears of most other opossums.

minutes to six hours, and one explanation for its origins, evolutionarily, is that it began as a form of intraspecific submission (similar to the way one dog bows down and tucks its tail to reduce the risk of injury from a fiercer, more aggressive dog). Instead of the posture saying, "Don't hurt me, I give in," this behavior says something more like, "Don't hurt me, I'm so non-threatening that I don't even exist anymore."

For our next forest-floor animal, we turn from mammals to birds, specifically to the green ibis of Central and South America. Well, it *looks* good, anyway, with its blue legs and glossy green neck and back. More information than "it looks nice" is hard to come by. Here is what Cornell University's *Birds of the World* website has to say: "Very little is known about some fundamental aspects of the natural history of the green ibis, such as the diet and the reproductive biology." It *seems* they eat worms, insects, fish, and maybe frogs, but how much of each and of what size and in what ratio all remains to be cataloged. The birding site eBird calls it "dumpy," which seems not very nice, though the site does concede that this ibis shows a "shimmering emerald and bronzy iridescence that can be stunning in good light." In parts of its range, it competes with up to six other species of ibis—glossy, white, scarlet, and so on—and while those inhabit more open, marshy habitat, the green ibis stays in the forest, walking alone and probing with its sensitive beak. It can detect very minor vibrations in the water or mud, and a special part of the ibis's brain processes this sensory input with blazing, M-chip speed.

Our green ibis photo on the next page is a camera-trap image from Nicaragua; in essence, the bird took its own picture by triggering a motion sensor at dusk and creating an unintentional self-portrait. Under the heading of SOS, or "science on a shoe-string," my photo partner, José Gabriel, and I have set up several different passive camera traps. (Camera traps are a remote field-observation tactic in which infrared-triggered cameras are installed and left in place by observers to capture images in their absence.) Some units we leave in the field for months before we have a chance to come back. Each trap we install shares something in common with all the others: it uses a second-hand camera, hacked battery packs, whichever lens we use the least often, and camera and flash housings that have been made from everyday materials like PVC pipe and duct tape.

As it turns out, you can do more birding with camera traps than people might suppose. As we go to press, the California Bird Records Committee has released the most recently updated list of accepted and official California birds. From American avocet to Xantus's murrelet, 686 species have been documented, including native birds, wild vagrants, and self-sustaining populations of introduced groups like parrots and wild turkeys.

The most recent addition to the California list was the Siberian rubythroat. This Old World flycatcher breeds in Russia and Mongolia and winters in Southeast Asia. It was documented as being in California in November, when it should have been in

Myanmar, not Mountain View. It would have made diehard bird
listers deliriously happy if they had been able to see it, except no
humans ever encountered it. The only trace of its passage was
a single "selfie" taken with a trail camera on campus at Google
headquarters. By the time somebody checked the memory card,
it was too late. The bird had long since left.

Defensive Strategies

If you live high in the trees or spend most of your life like a mole,
burrowing underground (a "fossorial" lifestyle), then the number of things that can catch you and eat you are proportionally
reduced. The same is true if you usually fly swiftly or can swim
with warp-speed acceleration. But walking around on the forest
floor noshing on palm nuts? That's rough, since everything from
owls to snakes to jaguars might want to have a go at you.

One obvious defense is to coat yourself in sharp quills. There
are thirty species of porcupines worldwide, ranging in size from
species that can be as small as two pounds all the way up to the
big lumbering tanks of Africa, which can weigh sixty pounds.
Romans used to hunt porcupines, which is how African porcupines got introduced to Italy, where they still live today. Some

porcupines in Central and South America have prehensile tails, including the one shown on the next page, the Quichua or Andean porcupine. These prehensile-tailed species can dangle from tree limbs to reach fruit or investigate an interesting knothole. They may be asking, "Why should monkeys have all the fun?"

In the desert chapter, we talked about convergent evolution. Animals in very different places and with very different ancestry evolve in parallel ways, "converging" on a common body plan. Another example of that is the hedgehog cohort and several hedgehog look-alikes.

While the European hedgehog is famous from medieval bestiaries and other folkloric sources, there is more than just that one species. As a group, hedgehogs are diverse and widespread. Seventeen species can be found across the globe, living in countries as varied as Scotland, India, China, Iraq, and Somalia. There are no hedgehogs in the New World unless you count pets, but non-native European hedgehogs are now feral pests in New Zealand. They were released in 1870 in an attempt to make New Zealand more British. It was a misguided impulse, since hedgehogs harm native species. Eradication campaigns intended to undo the damage are ongoing.

The Daurian hedgehog lives only in southern Russia (the Transbaikal region, or "Dauria") and northern Mongolia. Some sources say that hedgehogs have spines, not quills, though the distinction is not universally made. Call their spikes what you will, what is true is that unlike porcupines, hedgehogs can curl up into an impenetrable ball. I've held them when they are in their defense mode, and it feels like holding a large, hefty grapefruit that has been covered all over with high-quality plastic skewers. Hedgehog spines don't draw blood at first touch—they are not that sharp—but you wouldn't want a face full of them, that's for sure.

In the southern hemisphere, the hedgehog look-alikes are called echidnas, and the four living echidna species are each so odd and so ancient that all of them still lay eggs. The platypus does as well, but no other mammal besides those two groups does. Besides being egg-laying, echidnas have another bird trait. Just like a barnyard hen, both sexes of echidnas have a single, multifunction orifice, the cloaca, which is used for urination, defecation, and mating.

As is true with hedgehogs and tenrecs, the echidna's spines are made of keratin, the same miracle protein that also creates hair, wool, fingernails, feathers, horns, beaks, claws, and horse hooves. Echidnas are sometimes called spiny anteaters, but the true anteaters are all in the American tropics, while echidnas only live in Australia and New Guinea. In any case, the western long-beaked echidna of West Papua, eats worms, not ants. Somehow, I don't think calling it the "spiny worm-eater" will catch on, but in the Americas there is a snappy-looking passerine called the worm-eating warbler. People don't seem to mind its name, so anything is possible.

We turn now to a final example of convergent hedgehoggery. Madagascar is a large island east of Africa that broke off in stages. When the supercontinent of Gondwana broke up, what later became Madagascar and India first split from Africa around 135 million years ago. Then that land mass split further, and Madagascar as we know it separated about 88 million years

▼ Quichua porcupines live in South America and have prehensile tails that can hold onto a branch the way a monkey's tail can.

▲ Daurian hedgehogs live in forests and open steppes in Russia and Mongolia.

▲ The western long-beaked echidna is a nocturnal, egg-laying mammal related to the platypus. Young echidnas are called puggles.

◀ Here is convergent evolution in action: A lesser hedgehog tenrec explores a forest in Madagascar.

ago. That long gift of time has granted the native fauna many chances to become more "Madagascar-an" than African. This is the home island for lemurs, for example, which we will meet in the night monkey chapter, while birders come to see vangas and cuckoo-rollers.

Mammals called tenrecs have evolved here to be very, very hedgehog-ish, without being related at all to the official hedgehog lineage. They look like hedgehogs, they act like hedgehogs, and for all I know, they even smell like them, but according to the systematic lists, Madagascar's tenrecs' closest relatives are otter shrews and golden moles. With Madagascar's wildlife, we are off in a very special cul-de-sac of evolution.

Tenrecs are nocturnal and eat insects just like regular hedgehogs, and they go into a mild hibernation during winter. They are starting to become popular in the exotic pet trade. This picture, though, was taken in the wild the old-fashioned way: by stretching out prone over a bed of thorny sticks on top of a nest of Dracula ants just as it began to rain.

The Phuket pricklenape wants to be a hedgehog but can't, since it is a lizard, but it puts on a spine show all its own. Reptile scales are made from keratin, too, but there are different sub-types, so this agamid lizard is covered in beta-keratins, but human skin is made from alpha-keratins. Only identified as a distinct species in 2015, the Phuket lizard is a Thailand endemic that shows us that punk ain't dead, it's just coming back for another revival.

Some animal species are stable and well-known; we are unlikely to learn about any new kinds of bears, and all the great apes have probably been discovered. In contrast, the Asian pricklenapes are still expanding as a group. After the Phuket prick-lenape was raised to species level, other new pricklenapes were identified in 2018, 2019, 2020, and 2022. (Perhaps Covid slowed work in 2021.) At that rate of investigation, expect even more new lizard species by the time this book is printed and on shelves.

Another name for this lizard's genus is "horned dragon." With the Phuket lizard, I particularly like the two billy goat horns on the very front of the head past the main row of spines. Give it a pitchfork and a red cape and it would make the cutest little Halloween devil.

There are other ways to warn off enemies besides spines, and color is one of them. "Vivid by night, cryptic by day" was the reply when I asked a herpetology fan about the color pattern of Wagler's pit viper. This gorgeous but deadly reptile, also called the temple snake and the bamboo snake, is found in lowland forest in Southeast Asia. The "pit" in the name refers to heat-sensing glands on either side of the triangular head. Bright as this snake is, color patterns do vary; some individual snakes are more sedately white underneath, rather than the chrome yellow shown on page 158. The colors do help it hide, even if that seems improbable. On their first trips to the tropics, novice birdwatchers are often astounded at how well parrots can blend into the canopy. In the zoo or pet shop (or nature documentary), parrots seem obvious: Many are bright green with red heads and some even have red in their wings. They squawk and bicker and move around in noisy gangs. How can one *not* see a parrot?

It turns out that bright green with a splash of red matches rainforest foliage all too well. As the first shot in this chapter

showed, when you look up into the canopy, it all becomes a shifting mosaic of light and dark. Where does one tree end and another begin? There are random flowers and dead leaves and lots of backlit shapes. Some of the leaves themselves are so big that a full-grown parrot is completely hidden behind them. In the field, even though you can *hear* the parrots, and even though you saw a flock of parrots fly into a given tree, it turns out that *seeing* them is another task altogether.

With the Wagler's pit viper and many others, the head pattern shows a racing stripe across the face. That is because the line through the eye breaks up the silhouette, making it look not like a living creature but just another stick. In daylight the yellow (or white) underside creates a reflective surface for what is called countershading. Imagine a generic sandpiper, white below and mud-tan above. If the bird is standing in the open sun, the harsh light washes out the upper part's tan, helping it match the color of the substrate. Meanwhile, that same light hits the mud and bounces back up, so a white belly looks darker than it really is. Through countershading, the top and bottom halves end up blending into a mud-colored blob, hard to pick out from all the other mud and debris on the beach or mudflat. (Modern beaches are often cleaner and less jetsam covered than they were in the past; piles of seaweed and driftwood were the historical norm,

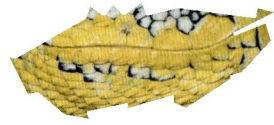

A camera's flash captures the vivid colors of a Wagler's pit viper.

providing more food and cover for wildlife.) On this snake, the yellow does the same thing in terms of being a reflective surface for all the green light bouncing around the interior of a forest. Bands of light and dark break up the long shape as well. If you do notice it, the viper's colors warn you to be cautious: This one bites, and it packs some serious venom. "Hands off, buddy. One step more and it will be the last step you ever take."

At night, of course, the colors do not matter; they only show up when a photographer with a Nikon camera and a flashlight documents the scene. This is such a beautiful snake I am glad somebody took the risk to document it for our book.

How to Catch Dinner

So much for the hunted, but what about the forest floor's hunters?

One point to remember is that few aspects of nature are ever fully compartmentalized. So, while we tend to think of pumas as the iconic ghost cats of the American West, they also occur in swamps and rainforests, and they range from Alaska through Central America and the Amazon, all the way south to the cold steppes of Patagonia.

Bobcats, too, can be rainforest hunters. Their southern range ends in Oaxaca, so they do not get into the tropical forests of the Yucatan or Costa Rica. (In those habitats, ocelots and margays fill the bobcat's niche.) And it is true, in places like Death Valley or

Point Reyes National Seashore, bobcats are open-country predators. In fact, Point Reyes is one of the best places in the world to
see a bobcat in the wild. Yet in the American Northwest, in parts
of Canada, and in the southern United States, bobcats are wood-
land animals, not desert ones.

Our photo below shows a young bobcat on the edge of a stream
in the Hoh Rain Forest, in Washington state's Olympic National
Park. Parts of this park get 200 inches of rain per year (and other
parts "only" 140 inches). This shot reminds us how "forest-floor-
ish" this species can be. Voles, quail, frogs, and fish—an alert bobcat
can find plenty of things to hunt in this lush, coastal rainforest. In
fall (just like the bears in Alaska), bobcats in Olympic National
Park specialize in seeking out spawning salmon. Bears and bobcats
alike eat their catch away from the water, inside the forest edge.
The salmon carcasses decompose and add significant amounts of
nitrogen to the soil. It may be an oversimplification, but one can
say that spring's flowers started out as fall's dead fish.

We think of predators as being large, fierce animals like tigers,
sharks, or harpy eagles, which are large enough to capture sloths
and monkeys. Yet the chase between hunters and hunted happens at all ecological scales, including down among the spiders
of the forest floor.

A young bob-
cat pauses at a
stream's edge
in Olympic
National Park.

Tarantulas, like scorpions, carry the burden of that "ewww" reputation, in which they are feared out of ignorance, often by people who have never even seen one in nature. Is it just that they are more plus-size than what we think of as the "proper" spider? Is it that they are spiders in the first place? Maybe it's the hairy legs? Something about even the idea of a tarantula triggers the "scream and grab a flamethrower" response from otherwise calm, rational people.

Is this the part of the book where we say, "Just deal with it, people"? There are a thousand tarantula species worldwide, so we may as well get used to the idea. Odds are they will be around long after *Homo sapiens* has joined the runners-up list in the record book of history.

Tarantulas are nocturnal hunters and are sensitive to vibrations; they can "feel" their prey nearby, almost as if (like the Marvel character Spider-Man) their spidey senses are tingling. Tarantulas hunt insects, other spiders, and, once in a while, lizards and frogs (small ones, anyway). Their hunting style is basically grab, inject venom, and finish the deed with the fangs. It sounds harsh, but they can't kill a person—their venom is too weak—though in one case in many thousands, an allergic reaction is possible. As with so many other things, don't mess with them and you can't get bitten in the first place. The largest spider in the world is a tarantula, but that does not mean much; the average species is a few inches across and is just doing what it is programed to do, which is to explore the forest floor and find things to eat.

While this book was in production, a new species of tarantula from Thailand was announced. It lives in mangroves, and in some light, its legs shine bright blue. As is also true for blue jays, the blue in this case is structural, due to the way the spider's hairs refract light. It is not actual, in-the-tissue pigment. Even so, blue is blue, and this new beast looks absolutely fabulous. Now I have spider envy. Why are all the tarantulas around

A tarantula in the Pantanal of Brazil blends in with the forest floor. Some tarantulas can live for thirty years.

me such a drab shade of dirty brown? I feel left out. America needs its own bright blue spiders. Maybe they are out there and we just haven't found them. Arachnologists, try harder, please!

Another forest predator has a good kill rate but a limited attack radius. In fact, it hunts not by going out on patrol itself, but by waiting in one place and luring potential prey to come to it. The giant pitcher plant of Borneo has special urn-shaped leaves that store both a large quantity of an insect-attracting nectar and also digestive fluid that allows the plant to compensate for mineral-poor soil by "eating" ants, flies, and other prey. All carnivorous plants grow in locations where the soil is poor in minerals and often very acidic. The plants compensate by attracting insects (and, very rarely, mice and birds), digesting them, and then gaining access to essential minerals like nitrogen and phosphorus.

The pitcher plant's method is simple. The primary well has slick sides; once they're in, the insects cannot climb back out. Only found on Mount Kinabalu, the plant shown in the picture on page 163 is endangered in the wild, but it is now cultivated widely by horticulturists. Those growing them at home can hand-insert insects into their captive plant, perhaps giving a short warning lecture to all the other flies in the room, or else they can use fish-food pellets.

A sister species to this plant attracts tree shrews—but not to kill them. The animals drink the nectar, and in trade, they defecate in the plant and fertilize it. That species has the main bell that is less slippery along the top edge than the typical pitcher plant, so the tree shrews can drink but not fall in.

This "strange but true" fact takes us to a final aspect of forest-floor ecology.

Strange but True

Were you one of those childhood nerds who loved to go to the public library and check out all the science books on the shelf? Perhaps you still are, and if so, right on and Godspeed. People who love books will never be bored and never be lonely.

As a kid, I always liked those "strange but true" parts of nature, such as the characteristics of the pitcher plant or the assassin bug (discussed on the next page). Books helped me imagine a world beyond my own small neighborhood. We can't finish up a tour of forests after dark without touching on three examples of "how cool is that?" If you're a super-hardcore nerd, you may know about at least one of these, but for most people, these stories will be new.

We start with the idea that some snakes live mostly in trees, some snakes live in saltwater (the sea snakes of tropical oceans), some live on the forest floor, and some even can glide tree to tree, like the flying snake of Borneo. Less commonly discussed are fossorial snakes, the burrowing kind that spend time under the leaf litter or in the top layers of rich, tropical soil. Making a rare aboveground appearance, the Costa Rican earth snake is the blue serpent shown on the next page on top of large brown leaves. I like the look of it myself, but the shiny scales are probably related to being able to move easily under the leaf litter rather than being useful as a signal or intended to reward our human sense of beauty.

The Costa Rican earth snake is found from Honduras to Panama; one name in Spanish is *culebra minera*, or the "mining snake." In the rainy season, they are on the surface more often than usual, but even then, they move around only at night.

Here is where the story gets interesting. This snake has a sister species, *Geophis dunni*, that is known only from a single specimen. A researcher collected coral snakes (another fossorial species) in the early 1930s in Nicaragua and took the preserved

▲ This pitcher plant in Borneo lures insects into its well. In time, the plant digests them, the nutrients making up for any deficiencies in the soil.

◀ This Costa Rican earth snake is attractive but rarely seen; it spends its life under the forest floor, burrowing in leaf litter.

examples to Germany. A few years later, a student was studying the diets of coral snakes and started opening the stomachs of the specimens that had been collected on that trip. Some snakes eat other snakes, which they have to swallow whole; a snake can't chew things up like a wolf or bear can. Inside a random coral snake, the student found *G. dunni*, and that one became the holotype, meaning the voucher specimen used to describe the new species formally. No one has ever seen *G. dunni* alive, and that dead specimen from the other snake's belly is still the only one that has ever been identified.

We turn now to something smaller than a snake, but in its own way just as deadly. Assassin bugs are long-legged predators from North and South America. They feed by inserting their beak into their prey, injecting digestive fluids, and sucking out the resulting goo. To help catch their prey, they secrete a sticky substance onto the hairs on their front legs, similar to what the sundew plant does. These sticky hairs help the bug to hold onto prey and can even snag small bugs flying past. The assassin bug eats anything smaller than it is, including insect eggs, aphids, weevils, and leafhoppers.

Do names matter? Assassin bugs are part of a larger group that also includes the ambush bugs and the kissing bugs (because they bite you in your sleep), so the first thing they need to do is to unionize and hire a good publicist. It just sounds so grim, doesn't it—the idea of being liquified from within? We have science fiction tales about that, such as the 2013 movie *Under the Skin*, in which a space alien played by Scarlett Johansson ("the female" is her only title) picks up men in bars and lures them into a mysterious pool that liquifies them. To be dissolved seems somehow worse than merely being eaten, yet spiders feed this way all the time. Is it any different from our own tastes? As many critics have pointed out, our own traditions, from lobster festivals to beer pong to pie-eating contests, show that humans have our own bizarre food rituals. Anybody studying us who was not us might find themselves scratching their heads a lot.

Next up, we consider a strange animal with a strange name. Most people are familiar with the basic A-list animals like foxes or zebras, and some people know the names of the B-list species, such as caribou, wild boar, or honey badger. Beyond those, though, most social awareness drops off quickly, so only a very dedicated mammal fancier (or an Australian conservation officer) has ever heard of our final strange-but-true entry, the northern quoll.

A quoll is a cat-size carnivorous marsupial whose name comes from Captain Cook's mistranscription of an Aboriginal word. There are two quoll species in New Guinea and four quolls in Australia: tiger, eastern, western, and northern. To me they look like a cross between a spotted skunk and an elephant shrew; other comparisons are to ferrets, or to a slim, mean opossum. Either

Northern quolls live in Australia; all males die after mating.

way, all quolls are nocturnal. They eat birds, mammals, fruit, crabs, insects, and lizards, only the problem is that some quolls also eat cane toads, a poisonous non-native species originally from South America. The toxins are so strong and the toads are so large that a quoll can die after eating only one cane toad (see page 292). Quolls also get hit by cars and are attacked by feral dogs and cats, so that in Australia there is concern about the best ways to preserve them and make sure they do not become extinct.

That is not the quolls' only worry.

In the ultimate example of truth being stranger than fiction, we open now to page 48, *A Field Guide to the Mammals of Australia*, second edition. For the northern quoll entry, the ending sentence states matter-of-factly, "All males die after mating." Wait—*what*? No more information is given. This mortality is not due to any action on the female's part—she is not an Aussie version of a praying mantis or black widow—nor is it misguided self-sacrifice to leave more fruit, insects, and cane toads for everybody else. A quoll certainly does not have the Lone Ranger's level of self-awareness to dust off their hands and say, "It looks like my work here is done."

Current research indicates that male northern quolls work themselves to death in a frantic, sleep-deprived, nonstop month of battling rivals and seeking out as many lady quolls as they can. Stress kills, as we all know, and in their case, the expression is literal. They focus so intently on the urge to pass on their genes that all else becomes secondary, even food, grooming, and sleep. They burn the candle at both ends, but as the original poem claims, that is all right, because it gives off such a lovely light.

To all male quolls past, present, and future, we can only say that we hope it was worth it.

Pumas, Ocelots, and Kodkods: A Cornucopia of Cats

There are forty species of wild cats ("felids") in the world, and so far, no one person has seen them all in the wild. A few of the hardcore mammal spotters are getting close, though, and the current frontrunner, Phil Telfer from Britain, provided the oncilla shot on page 183 (right). That photograph reveals either an admirable dose of passion and dedication or that all Brits are barking mad, since Phil flew from London to Colombia for the weekend, just for the chance to see that one cat. He had missed that species before, but this time he got lucky. That sighting put his cat list at thirty-six—only four more species to go. He admits that it could take another five to ten years to check off those last four (although he hopes to achieve it sooner).

Wild cats fascinate us the way no other animals can. In the past, all circuses had to have a lion tamer and big cats, and even now trophy hunters still want to kill—and display—their prize carnivores. Two dozen countries have issued snow leopard postage stamps, and there

Pumas have been cool for as long as there have been people in the Americas. This petro-glyph is on display in Pet-rified Forest National Park, Arizona.

are jaguar sports cars, *Lion King* musicals, sexy dancing tigers in the movie *Zootopia*, and a leopard-skin pillbox hat immortalized by Bob Dylan. His most dedicated fans are called Bobcats. But has Mr. Dylan ever flown in the Puma company's private jet? Sorry, Bob, but Jay-Z has.

Here is the formula all cats share; it's a successful, eternal design: Start with a lithe and flexible body, one that is deeply but not extravagantly muscular. Add four springy legs and an expressive tail, usually (but not always) long. Retractable claws. Binocular vision with eyes that can see wickedly well in the dark. (How good is their night vision? Six times better than yours.) Keen hearing, especially the lesser cats—they can hear a scur-rying rodent nearly as well as an owl can. Add whiskers, which are useful for navigating narrow places, and raspy, bone-cleaning tongues. A special part of a cat's mouth tastes the air—the Jacobson's organ—so they can smell another cat (especially one in heat), take in microparticles of their prey (as scent) while hunting, or pass judgment about a hunter's cologne. Cats can hiss, mew, spit, chuff, and growl. Some roar; some (like an in-the-mood puma) make a screaming sort of *kree-oww* sound. Most wild cats are born with spots; many lose these later on. Black phases are possible, including for jaguars, leopards, servals, oncillas, Geoffroy's cats, pampas cats, bobcats, and Asian golden cats. Black cats are more typical in rainforests, and among other advantages, their coloring may help control parasites. Cave art in France captures the shape and color of now-extinct cave lions. (And cave art also shows leopards, which are often misidentified as being snow leopards. Snow leopards were not present in Europe in the Pleistocene; these drawings show an Ice Age version of the regular leopard.)

What is a cat? Literature gives us a good answer in this excerpt from *Jubilate Agno* by Christopher Smart, 1759:

For I will consider my Cat Jeoffry.
For he is the servant of the Living God duly and daily serving him.
For at the first glance of the glory of God in the East he worships in his way.
For this is done by wreathing his body seven times round with elegant quickness.
For then he leaps up to catch the musk, which is the blessing of God upon his prayer.
For he rolls upon prank to work it in.
For having done duty and received blessing he begins to consider himself.
For this he performs in ten degrees.
For first he looks upon his forepaws to see if they are clean.
For secondly he kicks up behind to clear away there.
For thirdly he works it upon stretch with the forepaws extended.
For fourthly he sharpens his paws by wood.
For fifthly he washes himself.
For sixthly he rolls upon wash.
For seventhly he fleas himself, that he may not be interrupted upon the beat.
For eighthly he rubs himself against a post.
For ninthly he looks up for his instructions.
For tenthly he goes in quest of food.
For having consider'd God and himself he will [now] consider his neighbor.

That this text was generated by an inmate in a mental asylum does not lessen the accuracy of its perceptions.

Studying marks on the trail, you can tell a cat has passed by the shape of the tracks. Bears have footprints that look like human feet, and dog tracks show the small indentations of their non-retractable claws, but all cats' feet look like the paw in the photograph at right. The paw pictured belongs to a mountain lion, but the shape and the ratio of pad-to-fur can be seen in all the world's felines.

Cat paws all follow a similar template to this mountain lion's paw. The paw's owner has been darted to be fitted with a radio collar.

Male lions, of course, have manes, unless they are the Asiatic lion of India, a unique and mostly mane-less subspecies. Lions historically had a wide distribution. Around the start of the Roman Empire, lions ranged through Africa as they do today, but their range also extended across Greece, Palestine, Asia Minor, Iran, present-day Georgia, Pakistan, and much of India. Earlier than that, in the Bronze Age, lions occurred as far north as Spain and France. Will they ever be back? I hope so.

Lions hunt at night more than they do during the day, which in Africa is partly about heat and mostly about stealth. They also like to hunt at dusk, dawn, and during noisy rainstorms. Studies suggest that for attacks on humans by lions, the most dangerous time is during early evening for the week after a full moon. Lions are less successful at hunting during nights of bright moonlight, meaning that after such a period, they are hungry and hence more aggressive. Usually, lions have more to fear from us than the other way around, but being hunted and killed by lions remains a serious concern for rural people.

A full lion is, of course, less of a threat to humans than a hungry lion, and lions on the safari circuit (for better or worse) grow accustomed to the nuisance of the tourist trade. One night in Botswana, my jeep was parked ten feet from feeding lions. We turned the engine off and killed the lights and just listened. The lions didn't pay any attention to us; they knew we were there, but they were focused on the zebra they had taken down. Sitting in

A lion patrols after dark—a sight that would have made our primate ancestors very uneasy.

A lion patrols after dark—a sight that would have made our primate ancestors very uneasy.

the dark and hearing them crunch bones and gnaw joints did, I admit, give me a case of the primordial heebie-jeebies. Despite how advanced we may think we are, being close to feeding carnivores forces you to appreciate just where you rank on the overall food chain. We may be tough and tech savvy on the outside, but we are all scared primates on the inside.

The only cat that outranks lions on the pop-culture scale is the tiger. William Blake's "The Tyger" is supposedly the most anthologized poem in English. Richard Parker is either a real tiger, a figment of a castaway's imagination, a literary symbol, or none of the above in the novel and movie *The Life of Pi*. The Chinese zodiac has a Year of the Tiger. In Hindu mythology, the warrior goddess Durga rides a tiger. "Put a Tiger in Your Tank," ads for Esso brand gasoline exhorted me in my childhood. In *Apocalypse Now*, a tiger sighting powerfully reinforces the crew's rule: "Never get out of the boat."

A well-publicized subcategory of general tiger-ness is the idea of the Siberian tiger. Tiger sizes tend to increase from south to north, and pelts tend to be slightly paler the farther north the animals live. And while the exact size of any given tiger varies by location, sex, diet, and genetics, some monster Siberian tigers have been documented. One weighed almost five hundred pounds and, including its tail, was over ten feet long. Siberian tigers are the snow cats of the Russian Far East; their range is not actually in Siberia, but it's close enough to count for most people. Siberian tigers are never white with black stripes; the white tigers of Siegfried & Roy's magic acts were captive-bred Bengal tigers with blue eyes and recessive genes.

Like lions, tigers hunt by night more often than by day; unlike lions, tigers usually work alone. The target animal is killed by having its neck broken if it is small enough to be knocked over by the cat's initial lunge; if the prey is larger, the tiger goes for the throat and crushes the windpipe. Tigers in India typically eat sambar (large, elk-like deer), chital (another deer, but this one has white spots), Indian gazelles, and Eurasian wild boars. Necessity combined with opportunity adds termites, crabs, birds, turtles, fish, hares, snakes, and village dogs to that list. Yet a tiger can even take down a gaur, or Indian bison; this massive, oxlike beast is the largest wild cow in the world, with males weighing over one ton. If it moves and it is not an adult elephant, the tiger is ready to try to hunt it.

Sadly, there are more tigers in captivity than there are left in the wild. Even if zoo tigers could be trained and released, the

Some tigers measure more than ten feet long, nose to tail. They can take down a gaur, the two-thousand-pound Indian buffalo.

question is, "Where?" Habitat loss limits tiger populations in India and elsewhere, not low fecundity or any lack of hunting prowess. They need jungles to return to and game to hunt once they're there. If you want to save a tiger, plant a tree.

Four and a Half Lynxes and One Puma

There are four lynxes in the world—or more like four and a half, really, but we will get to that in a moment. The Canadian lynx, shown below, lives in Alaska, Canada, and northern-tier US states like Minnesota and Maine. Several hundred reintroduced lynxes live in Colorado but are rarely seen.

As with most wild cats, photos of lynxes usually show them in daylight. That's because those are the hours when zoos are open and also the time when most wildlife photographers are out and about in Denali National Park in Alaska or driving winter roads in Minnesota. (Winter in Minnesota can be good for lynx spotting, such as along the Gunflint Trail. Mammalwatching.com has trip reports with specific routes.) Doing photography during human hours, not cat hours, skews the results. If you look up

Grayer, larger, and more northern than a bobcat, a Canadian lynx has large, snowshoelike paws.

"lynx" online, seeing so many daytime shots might make it seem that these cats are more diurnal than they really are. Yet we know from trail cameras and tracking collars that the typical lynx hunts more at night than it does during the day. It's just hard to show that. For example, when I saw a lynx in Denali at four in the morning, it was too dark to get a picture, plus it didn't stick around. That was in May, when that part of Alaska still has darkness for a few hours in the middle of the night. Winter may make cats more visible midday: Their prey may be taking advantage of the (slightly) warmer temperatures, plus predators have to hunt more hours overall in order to survive. The Canadian lynx specializes in hunting the snowshoe hare.

Historically, this lynx was trapped for its fur, and as recently as 1986, you could buy a Canadian lynx fur coat—real fur, not synthetic—in upscale New York department stores and boutique furriers. The coat's price was $30,000, at a time when a new Cadillac cost $20,000 and a schoolteacher made $26,000 per year. Tastes change. According to activists' slogans, the only thing the fur coat looks good on is the animal itself.

North America's other lynx is the bobcat, named for its abbreviated ("bobbed") tail. We met it first in the forest chapter, page 159. Europe has two lynxes as well: the Iberian lynx in Spain and the Eurasian lynx, whose range stretches east from Scandinavia all the way to Russia, Mongolia, and China. Both of them look like the Canadian lynx, except the Eurasian one is larger (it can eat deer in addition to rabbits). The Iberian lynx used to be called the rarest cat in the world because so few were left, but reintroduction programs have had encouraging success. It is not that hard to catch a glimpse of one now; wildlife tours have itineraries that focus on this species, and mammalwatching.com has lots of tips.

Found from the Yukon to Tierra del Fuego, the mountain lion (aka puma or cougar) is a tan, long-legged, relatively small-headed apex predator. Puma kittens have turquoise-blue

MOUNTAIN LION HABITAT

YOU ARE IN
MOUNTAIN LION COUNTRY
If you encounter a lion:
DO NOT approach
DO NOT crouch down
DO NOT make sudden movements
Stand tall and make yourself appear as big as possible

A sign near a US Forest Service ranger station gives good advice.

eyes when young that will change to amber over time; they also have spots that soon fade. The mountain lion is documented with trail cameras way more often than it is seen in real time, though in Chile, around Torres del Paine, there is a growing puma-watching industry. In the American West, sightings are always cases of random luck, such as when one of these cats crosses the road right in front of your car or watches you from a far ridge before disappearing down the other side. I saw a puma in the Everglades once, and when I reported it at the visitor center, the ranger told me they had worked there for thirty years and never seen one. Then again, I have never seen a mountain lion anywhere else in North America, while my nature friends have all seen one or more.

If you do see a puma on the trail, be loud, act tall, and above all, do not run. The sign shown in the photo above covers the basics.

Mountain lions typically hunt deer or the deer-equivalent, such as a llama-like camel in Patagonia called the guanaco. The famous mountain lion of Los Angeles, P-22, did eat deer, but it also ate coyotes and Virginia opossums, and once, when it broke into the Los Angeles Zoo, an Australian koala. In the neotropics, where deer (and koalas) are less abundant, mountain lions hunt pacas and agoutis, peccaries, iguanas, rabbits, and the tropical alligator called a caiman.

What's in a name? Although *mountain lion* remains the more official, high-end term—the label one might see on a sign in a museum or zoo—I have noticed that among hikers and field biologists, *puma* is used more often in daily conversation. Perhaps it is just a punchier word, shorter and more direct? The term originally comes from the Quechua language, as do our modern English words for lima bean, bat guano, Andean condors, and the malaria treatment quinine. In the name of the Hollywood cat cited above, P-22, the "P" stands for puma. As another supposed *puma* synonym, *catamount* still gets listed in reference books, but I've never heard anyone use it, other than ironically. Ranchers in the American West might use *cougar* a bit more often than field biologists do, but all three terms remain interchangeable and socially legible, group to group to group.

The hills are alive with the sound of music, or so Julie Andrews tells us. They also are alive with the sound of camera shutters clicking all night long, from the trail cameras placed in just about every US national park and nature reserve. The black-and-white mountain lion shot that follows is from Golden Gate National Recreation Area. From where that shot was taken, it is only eight miles in a straight line to SFO, San Francisco International Airport. That style of camera uses an infrared sensor and invisible infrared flashes that read an animal's heat signature. (It also would capture any passing coyotes, bobcats, opossums, raccoons, and midnight trail runners.) Prices for this equipment drop each year, it seems, while image quality improves; this is a simple tool to buy if you want to survey your own regional trails. Some housings come with cable loops to deter theft, and one of my friends has made up labels explaining to passersby that the camera is for wildlife study, not hunting, which in his area is an unpopular sport.

The color trail-camera shot used a conventional flash trig-gered by a motion sensor. As the cat breaks the beam, the flash

Trail cameras near San Francisco and in the Montezuma Castle National Monument park document nocturnal pumas in both infrared and color.

fires. It is a more intrusive method but lets the researchers see more details. The shot above from Arizona has captured a heavy-bodied, well-fed cat, and this image verifies that, at least for this animal, things are going well indeed. Most mountain lions want to feed on a deer-size prey item about once a week, and this cat has clearly been having good success.

Dots, Spots, Swirls, and Curls: More Rainforest Cats

Jaguars are the tigers of the New World. Heavier than pumas and faster than horses, they can take down a tapir, crack open a huge Amazon turtle, poach a steer, and swim across the widest river. Maybe they would not do all of that in one day, but the potential is there. And even though tigers are larger than jaguars, jaguars have stronger jaws, with a bite force of 1500 pounds per square inch. That is twice the psi of tigers and ten times greater than what a puny human mouth can achieve. Rather than crushing the windpipe, jaguars kill their prey with a bite to the back of the skull, severing the spine.

While the cat known to many in the western United States as *el tigre* is usually thought of as a jungle-only cat, jaguars were (and are) part of the North American faunal assemblage, with eighty-plus historic and modern jaguar records in Arizona alone.

A jaguar demonstrates what would likely be the last thing you saw if you were an agouti and the jaguar was hangry.

One modern record comes from a border-patrol officer who saw a jaguar from his airplane. Based on specimens, photographs, letters, and newspaper articles, it is now clear that jaguars in Arizona were widespread, and since Anglo arrival, jaguars have been sighted (and persecuted) from the Mexican border all the way to the Grand Canyon. These sightings included females with cubs, indicating that jaguars were breeding in the United States up until the 1940s. They also inhabited California in the nineteenth century, coexisting with Mexican wolves, California condors, mountain lions, and grizzly bears. Do you ever wish you had been born in another time? The jaguars in Arizona and California were hunting deer but also everything from pronghorn antelope to elk to bighorn sheep. In addition, in North America, jaguars have been documented killing burros, horses, frogs, and at least one black bear.

With jaguars, if the question is "What do they eat?" the correct answer is "Anything they want to eat."

After the jaguar, the next best-known New World jungle cat is the ocelot. Lynx-size with a medium-length tail, it, too (like the jaguar), turns up in Arizona and extends south through Mexico, Central America and the Amazon Basin, and as far as Argentina. In the United States, besides the strays in Arizona, the only regular populations are at Laguna Atascosa National Wildlife Refuge and the adjacent Lennox Woods Preserve, both in south Texas.

The ocelot is one of several felid species that take the idea of spots to their most beautiful extreme. Like those of the marbled cat and clouded leopard of Asia and the king cheetah of Africa, the ocelot's markings flows across its shoulders like marbled chocolate. The Nature Conservancy compares it to "op art," noting that the pattern "consists of chainlike streaks, spots, blotches, and rosettes of dark markings." It is the largest of the small cats in the Americas, with a tail that nearly reaches—but does not quite reach—the ground.

Ocelots hunt at night unless it is very cloudy, in which case they may venture out during the day. Depending on where they live, they mostly eat rats, rabbits, opossums, iguanas, land crabs, and armadillos. If you live in North America, you're used to encountering only one kind of opossum: *Didelphis virginiana*, the Virginia opossum. In contrast, in a country like Panama there are ten opossum and mouse-opossum species, two kinds of armadillo, three sloths, and three species of anteater (see page 298). This is relevant not just because diversity and abundance are interesting, but to show that as the smaller felids partition resources, there is rarely just one thing (such as "an" anteater, singular) that they are competing for. All ten opossum species

Even at a fast trot, an ocelot still looks like an ocelot. Note the medium-length tail and the marbled stripes at the neck and shoulder.

would never be in one place, but this kind of diversity does mean that in almost all pieces of habitat in Panama, each ecological sub-type—from town dump to cloud forest to tidal swamp—will have one or more resident species.

Restoration ecologists and environmental activists use terms like *connectivity*. The ocelot is one reason why that concept keeps coming up. While the range map makes it seem like ocelots occur broadly and generally, at a local scale, if you were to zoom in, their needs center on microhabitats—a dense thicket of thorn scrub alongside a seasonal arroyo, or the narrow band of mangrove cover in between a tidal estuary and a busy highway. They do not just need prey to hunt (and a quiet, uninterrupted night in which to hunt it); during the day they need a place to rest. That site does not need to be anything elaborate, but it has to be safe, enclosed, and protected. In the daytime, ocelots might go up into the trees and sleep in dense vine tangles, or they may find an unused culvert under a road, a fallen log with a crawl space under it, or a tall, living tree with a hole in its trunk. They need drinking water and a den, so the female can give birth to her kittens and then stash them there safely while she is out hunting. (A mother ocelot with cubs to feed might be out hunting seventeen hours a day.) Ever wonder what the leading cause of death is for ocelots in Texas? Being hit by cars while trying to cross the road from one habitat patch to another.

Ocelots used to be sold as pets—I am old enough to have
seen this as a child—and this species, too (like the lynx), used
to be trapped and made into coats. Mink coats may be every bit
as unethical as cat-pelt coats, but all mink coats made in the
past hundred years have been from farm-raised animals, not
wild-caught ones. The term previously used was *ranched* mink,
as if the animals in question were allowed to gallop freely around
the sagebrush, kicking up their heels and chasing butterflies.
Mink farming is not that; it consists of cages and brutal effi-
ciency, but at least it does not involve steel traps and depleted
habitat. Until recently, it was normal and acceptable to wear
ocelot-fur coats, and I would guess in some clandestine circles
it might still be. For most of us, though, that kind of display is
unthinkable, and broadly speaking, society has moved on.

In zoos, ocelots live for up to twenty years. Field data is
sparse, but the best guess is that, in the wild, their life span is
closer to ten years. Ocelots are solitary and rarely vocalize in
the wild, communicating instead by scent marking. That trait
explains one way that researchers monitor the cats: If they douse
a post in men's cologne, the ocelot will investigate . . . and inadvertently take its own picture with a trail camera. As with the
stripe pattern on tigers or the barnacle patches on a whale's tale,
with a good side view, an ocelot's unique markings can be seen in
a photo, and then individual animals can be tracked with reliable
precision.

There are half a dozen recent records of ocelots from southeastern Arizona. Different species of night animals have different

kinds of reflections when you sweep the fields with a spotlight. For ocelots, the eyeshine is bright yellow, in case you ever decide to go look for it.

Good luck, and don't forget your camera.

Latin America's other small felids include the margay, which is like an ocelot but smaller and proportionally longer-tailed. It has elongated splotches on its coat as well, but they are never as distinctly oval as those seen on the ocelot. This cat climbs well and can even shimmy up a branch while hanging on upside down. Luke Hunter's *Field Guide to Carnivores of the World* calls it a "spectacularly acrobatic climber able to hunt the most agile prey." Besides opossums and agoutis, it eats shrews and birds.

If there were such a thing as a teacup-poodle version of a margay, it would be the tiger cat, or oncilla. Some live in Central America and northern South America, and a related population (related and yet different) lives in mainland South America. The northern tiger cat form has recently been split from the southern form; both are sometimes called tigrinas, and both sometimes oncillas, except sometimes the northern one is called the oncilla and the southern one is called the tigrina. When I first started reading about animals in the public library many years ago, there were fewer animals but more certainty. The usual pronunciation of *oncilla* is with the Spanish "y" sound for the doubled "l": *own-sea-yah*.

Geoffroy's cat, resident in scrub and forest in Uruguay, Bolivia, Brazil, and Argentina, is like a yellow housecat covered with many small, quick dabs of black paint from a deft, narrow brush. It is named after nineteenth-century French naturalist Étienne Geoffroy Saint-Hilaire, who admired Napoleon and worked for him, and whose name proves that back in the day, everybody had a better name than you do. Napoleon gave Geoffroy a medal for being very good at going to other countries and looting their art.

▲ An ocelot on Pipeline Road in Panama looks out of roadside foliage at an at-the-ready photographer.

◂ This northern oncilla in Colombia is part of a larger species complex. Multiple common names for the cat exist, but the animal itself doesn't care.

The Andean mountain cat lives at elevations of up to thirteen thousand feet in open, often desolate habitat. It is gray with rusty brown stripes and has thick fur but a slightly beat-up coloration, like it is trying to blend in with the corrugated tin of abandoned mining shacks. It eats viscachas, which are sort of like guinea pigs with rabbity ears.

And here we encounter some real taxonomic chaos: the Pantanal cat / Pampas cat / colocolo complex are brown and plain (or not-so-brown, not-so-plain), housecat-size, mid–South American felids that look enough like the Andean mountain cat that 50 percent of the posted sighting records are probably wrong. This complex of animals represents one, three, or six species, depending on which way the coin toss goes. For now, different specialists have different interpretations. Mammal watchers tend to go with the one that best fits their life list.

Kodkods—we mustn't leave out the kodkod—live all the way down at the very bottom of Chile and are one of my favorite species because (a) I have seen a kodkod, and (b) that name is just so fun to say. Some words have "mouth music," and that word is one of them. Another name for the kodkod is the güiña, and it is the smallest cat in the Americas: dinky and spotted and also available in the all-black, melanistic version. That form I have not seen; I guess I will need to go back to Chile.

There is a final plain-coated New World felid: the jaguarundi, which comes in a red phase or a slate gray phase and is shaped like a puma crossed with a weasel. They live from northern Mexico to Argentina, and one of the last sightings in the United States was in 1986, when one of my friends saw one near Brownsville, Texas. I have seen them in Mexico and Belize, both in the daytime and at night. According to Fiona Reid's *Field Guide to the Mammals of Central America & Southeast Mexico* (2nd edition), captive jaguarundis make "chirpy whistles and screechy noises."

The Clouded Leopard: Asia's Saber-Toothed Cat

Fans of the La Brea Tar Pits in Los Angeles are aware of the *Smilodon*, better known as the saber-toothed cat; it is the official state fossil of the state of California. Our usual narrative is that it went extinct at the end of the Pleistocene.

Of course it did . . . except in a way, it didn't. See the skull on page 186. Look at those teeth—wow, this is a modern-day saber-tooth if ever we had one.

The clouded leopard may have the most attractive coat of all of the cats.

The skull belongs to the clouded leopard from Asia. Technically, we have to say the clouded *leopards*, plural, since taxonomists have split them into two forms. One lives in Borneo and Sumatra and the other one in mainland Asia—Nepal, Bhutan, India, China, Malaysia, and so on. They look alike, and not many people remember to distinguish between them as separate species. So that they can update signage and labels, zookeepers and museum collection managers will need to figure out the origin locations for the leopard-y things in their care. For the rest of us, it's still all one thing, the magnificent, impossibly beautiful clouded leopard, singular.

The idea of a saber-toothed tiger—as we called it when I was a kid; it's saber-toothed *cat*, now—remains as exotic and thrilling today as it was when the first fossils began to be identified. The Latin name means (loosely) "fate scalpel tooth." Saberlike teeth evolved at least three times: in the true saber-cats, plus two lineages of "false" saber-cats, the nimravids and the barbourofelids. There is nothing so inherently "over the top" about these cats' large teeth that made their decline inevitable. Whatever killed them off—humans, climate shift, a change in prey behavior, or all of the above—it was not just their teeth being so very extra fang-some that the lineage was doomed by its own excess of dentition.

This was a widespread group. Even though the La Brea Tar Pits in California is the site most closely affiliated with the classic saber-tooth, *Smilodon fatalis*, fossils from this genus have also been found in Arkansas, Brazil, Florida, Indiana, Missouri, Nevada, Oklahoma, Peru, Tennessee, and Texas. *Smilodon* was one of the last of the machairodonts, an extinct branch of the cat family tree that evolved elongated, saberlike upper canine teeth. After crossing the Isthmus of Panama two million years ago, *Smilodon* evolved into two separate species: the North American *Smilodon fatalis* and the larger *Smilodon populator* in South America. Both species went extinct roughly ten thousand years ago, and the official cause of their demise is still unknown.

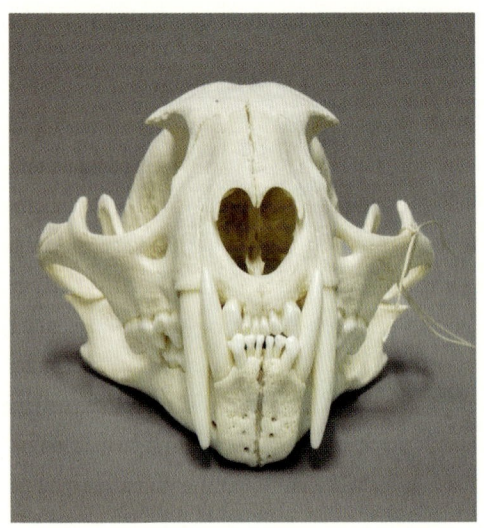

Smilodon's teeth are spectacular. How exactly did the saber-tooth cat deliver its killing bite? Some of the proposed answers include (a) *Smilodon* was a can-opener cat, with big teeth and a wide gape enabling it to cut through the tough, armored hides of giant sloths and huge armadillos; (b) *Smilodon* used its teeth to slash or stab; (c) It somehow punctured major arteries and lapped up the blood; (d) It ambushed prey and used its super-strong forearms and extra-strong neck to bite and hold on. This one makes sense: the skeleton clearly backs it up. But then what? (e) Either *Smilodon* would rip out the prey animal's throat, or (f) it would make a precise killing bite, severing the carotid artery. And option (g)? It could be something different from all these choices that has not been figured out yet.

Because La Brea has so many fossils of this one species, we can learn a lot about their daily lives by studying the bones. These animals had it rough, and their bones show evidence of arthritis, trauma, infection, and damage from the teeth of other saber-tooth cats. Was that from fighting or just getting tangled up in a feeding frenzy? Did they try to cannibalize each other? Nobody knows. Their own teeth could break off: Fractured saber-tooth teeth have been found embedded in the bones of prey animals. These cats also suffered broken legs, dislocated hips, bone infections, and back injuries. Long-term mechanical stress could cause microfractures, weakened bone, and bone thinning. Thirty percent of the one thousand skulls at La Brea show erosion of the parietal bone where the largest jaw muscles attached. Soft-tissue damage may have been just as severe, but we do not have the fossil data to show that.

The clouded leopard's extraordinary canines show why some people call it a modern-day saber-tooth cat.

And then, of course, there is the tar itself. The pools at La Brea might have been soft only seasonally. In winter, their texture might be like a freshly laid asphalt road—maybe a bit soft, with some blown-in twigs starting to stick, but the substrate would support your weight. In summer, La Brea's tar not only liquifies but perhaps builds up a "Judas layer" of twigs and fallen leaves right at the surface, making it look benign, potentially even solid. Under that skin of leaves and water hides a lethal layer of oozing tar.

The saber-tooth cat is still alive; we just call it the clouded leopard. In both the island form and the mainland one, they rest in trees during the day and hunt by night on the forest floor. The color pattern is exquisite; it takes the ocelot's coat and improves it, blending tans with grays and charcoal blacks. Despite the name, this is not a true leopard, and in fact, the two clouded leopard species are in their own genus—not a big cat, like a lion or tiger, and not one of the small cats, like ocelot or serval, but just its own midsize self. One folk name is "tree tiger," which is as good a fit as any.

The clouded leopard uses its tail for balancing when moving in trees and can climb down tree trunks headfirst. They predate pigs, monkeys, pangolins, civets, porcupines, birds, fish, and deer. In Borneo, they eat orangutans, which makes me frown a bit and ask rhetorically, "Oh, *must* you?" Similar to jaguars, clouded leopards are back-of-the-neck biters; the extra-long teeth allow them to remain small and light and yet have the lethality to take on a bearded pig or hog deer. For them, trees mean safety, since they can get up high enough that they won't encounter a tiger or elephant, to name two creatures of the forest floor who would outrank them in a fight. Once in the trees they can leap fifteen feet, which may not sound like much but is longer than the average SUV. No human can broad jump that far.

Clouded leopards also have odd eyes: The pupils never get fully round like a big cat's pupils do, yet they never shrink to vertical slits like a small cat's pupils do. Instead, they stay in an oblong shape. The San Diego Zoo says that it feeds its captive clouded leopards "a fortified meat-based carnivore diet" and encourages them to "gnaw on large beef knuckle bones." For variety (and to ward off boredom), captive felids are given blood popsicles and hanks of rope saturated with elk urine. San Diego Zoo's clouded leopards "sometimes eat chunks of papaya frozen in ice blocks." Standard zoo practices include giving their animals meat that still has skin, bones, or feathers—which is to say, better a dead chicken to pluck and mess around with than a bucket of chicken nuggets.

And Now, a Word About Man's Worst Best Friend

Fully feral cats combined with domestic cats that spend time outdoors are bad news for birds, lizards, and other small animals. It's not that the cats eat them, although they do that, too, but that they fatally injure them just as often. I don't want to seem unkind to the cat lovers in the world, but the numbers speak for themselves. One report in *Nature Communications* cited low-end estimates that owned cats (ones with homes but that are allowed to roam freely) and feral house cats (living full-time outdoors, but not originally native, like bobcats or pumas) cause 1.4 billion bird deaths annually and more than five billion mammal deaths. The bottom line? Domesticated cats belong in the house, not in nature.

Miguel Ordeñana is an expert in urban ecology who works at the Natural History Museum of Los Angeles County. Regarding outside cats, he says, "It seems like they are happy out there. And they likely are, but there are consequences. You're putting your cat at risk, there might be coyotes out in the neighborhood,

This feral cat in Oman was living near a nature reserve. How many birds had it already killed that week?

but you're also putting native birds and lizards at risk. Some are not adapted to defend themselves against cats. Even the most lethargic, fat cat is still a predator and can do a lot of damage to the ecosystem."

A Quick Note on Look-Alikes

Some animals look like cats and yet are not cats. Felid look-alikes are many, but these are some of the most famous and interesting. The fossa (pronounced "foo-suh") is sometimes called the puma of Madagascar. Though not a true cat—it evolved from a civet-like animal that arrived on the island 18 million years ago—it does look and hunt like a cat. Civets, mostly native to southeast Asia, have catlike characteristics but are not cats. The ringtail, though less catlike than fossas and civets, used to be called the "ring-tailed cat"—this cliff-nimble raccoon has a silky tail and a cat's curiosity.

CLOCKWISE FROM LEFT: A fossa on the prowl.

This Malayan civet in Borneo looks like a black-and-white ocelot with a bit of zebra mixed in.

A "ring-tailed cat" no longer, today this creature is just called "ring-tail."

Vampires, Bulldogs, and Free-Tails: A Bonanza of Bats

In the year 1250—in between the death of Genghis Khan and the start of the travels of Marco Polo, after the drafting of the Magna Carta and during one of the many Crusades to "liberate" the Middle East— an anonymous monk in an abbey in Northumberland drew two bats side by side on a sheet of parchment. These bats became part of a bestiary that ultimately included over one hundred real and imagined animals, from hydras to elephants to nightingales.

These two bats are my favorite drawings in all of medieval art. They look like a pair of eight-year-old kids in Halloween bat costumes, trying to look fierce. (They also remind me of the whimsical figures in *Where the Wild Things Are* by Maurice Sendak.) The impression of kids playing dress-up is reinforced by the bats' upright stance; real bats hang upside down, but these two tricksters look like they are wearing robes made out of pillowcases and holding up oven mitts as pincers.

Every group seems to create bat art. Some cultures carved bats into netsuke (those intricate Japanese button toggles), painted them on pottery, or, in the case of the Mayas, included them in their glyphs. There are field-guide-quality fruit bat illustrations from the Moghul empire in India and nightmare visions of bats in Goya's most disturbing and prophetic etchings, such as his print *The Sleep of Reason Produces Monsters*. One British artist, Monster Chetwynd, has a self-portrait of herself not as a crazy cat lady but as a crazy *bat* lady, with kohl-dark eyes and a wild bat headdress. She knows that even today, bats provoke strong reactions, much more so than would a hamster, a budgie, a corgi, or a water vole.

One misconception people have is the idea that there is just "bat," singular, as in the expression, "Oh look, I just saw a bat!" Yet bats are as diverse a group as rodents are, with 1472 species around the world. In fact, there are more kinds of bats than all the parrots, hummingbirds, eagles, herons, seagulls, and ducks put together.

According to fossil records—and there are not as many as we would like, of course—bats became bats early on and have stayed bats ever since. The fossil shown below comes from the Green River Formation, whose sediment layers present a continuous record of six million years, starting fifty million years ago. Lithology is the study of stone, and the lithology for these deposits reveals a blend of sandstone, mudstone, siltstone, limestone, coal,

➤ Two bats from the *Northumberland Bestiary*, published around 1250.

▼ This fossil bat from Wyoming is fifty million years old.

and volcanic ash. Fine sediment has preserved intricate details. Fossils found in the Green River Formation include the remains of stingrays, turtles, birds, fish, dragonflies, and this early kind of bat. According to analysis done by the staff at Fossil Butte National Monument, claws on each finger of this bat's wings indicate that it was probably an agile climber, perhaps crawling along the undersides of tree branches searching for insects.

We now know that relatively small changes in DNA can "flip the switch" and help an animal develop webbed limbs. Not all evolutionary changes are incremental. A sarcastic person might ask, "What good is half a wing?" But an early bat could have had a relatively large membrane from the start, and even if not, half a wing is better than no wing at all if avoiding a predator means letting go of the branch and plunge-gliding down to the ground.

Wings

All bats fly with their hands. Your hand and a bat's wing are more or less the same thing—they just stretch the bones out differently. Bat bones are thinner, as well. You have a thumb and a bat has a thumb; you have an elbow and a bat has an elbow. Inside a seal's flipper are the same bones as are inside a bat's wing, and a bat's wing has the same digits that there are inside a horse's leg. Nature is the ultimate recycler, and the same part—maybe made thicker or thinner each time, maybe moved back or stretched out—can manifest itself in a dozen different ways. From the pelvic girdle to the shoulder socket to our tiny human tailbone, a bat and a person are just two destinations on adjacent (but long-ago diverged) train tracks.

Related to this, bats are not mice with wings; the engine of evolution has taken bats on a different path than the one the rodents followed. If bats have a living relative at all, it would be the colugo of Asia, which is also called a flying lemur, even though it is not a primate. A colugo looks a bit like a flying squirrel, though it's not related closely to them, either. We're used to thinking in terms of side-by-side comparisons, so that a cheetah is a slender, super-fast leopard, or a zebra is an African kind of striped horse, even though in that case, it would be just as logical to say that a horse is a tall, unstriped, domesticated zebra. So then, what is a bat? It is a flying lemur except it is not a lemur, and even if it were, flying lemurs never fly—they glide.

The pale spear-nosed bat lives in Central and South America, and it is one of the night-pollinating, nectar-feeding bats. It also eats pollen directly, and it eats insects such as beetles and moths. While the one shown in the following photo is reddish brown, colors vary; some can be so dark they are nearly black, and others have a paler, more yellowish back and a white belly. Dominant males assemble up to a dozen females in a harem; other males live in bachelor clusters. They are vocal and add to their wide repertoire of sound with scent-based cues. Fiona Reid, the author of the previously mentioned guide to mammals, is equally talented as an illustrator and a naturalist and describes them as "muscular, velvety."

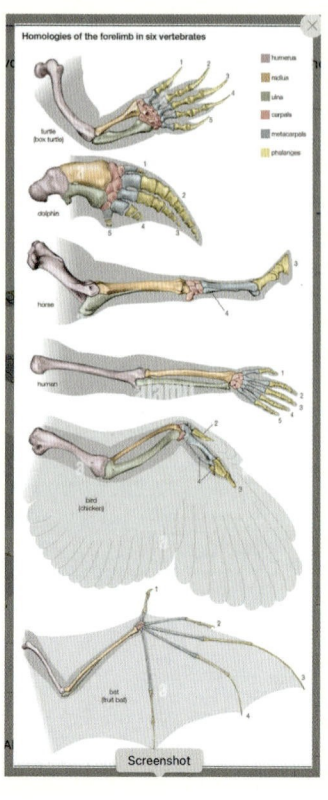

Our human hands are homologous with the limbs of other animals, including the wing bones of a bat.

Looking at the photograph of a spear-nosed bat on the next page, we can see that the bat has just completed a strong down-stroke, and in a fraction of a second it will va-voom clear out of the frame. The flash captured every detail, including the obvious thumbs at the top of each wing. When I was zoomed in on this shot during image prep, I could see a lot of small abrasions in the right-most wing, close to the body. If looked at closely, most bats' wing membranes do show shreds and punctures, in the same way that old jeans, no matter how well-made, eventually have frayed hems and worn butts. Bats forage in complex,

This is a pale spear-nosed bat from Central America. You can see its thumbs, one on the top of each wing.

skin-snagging environments that are full of thorny vines and leaves with serrated edges. Unlike blue jeans—but like our own papercuts and hiking blisters—their holes will heal in time.

Tails

Most bats have webbing around their legs and across their tails, for more lift when they are flying or (in the case of the bulldog bats) to help catch fish. The tail membrane may also help them scoop up insects. Some species have a long, visible tail that extends past the final membrane, such as the well-named mouse-tailed bats of the Old World or the Mexican free-tailed bat, a long-distance migrant we first met on page 54. Other bat species have the tail enclosed by the lower leg membranes, or else a tail that is so reduced it seems like no tail at all. Tails may help roosting bats that are wiggling butt-first up into tight crevices, or a tail could be an ancillary rudder in tight turns while a bat chases insects. Tails and leg-membranes may be negatives (because they add drag overall), and yet they can help, too. One study using wind tunnels and a model based on a bat's bone and membrane structures found that adjusting the model's tail position had a two-fold effect. The change increased camber—meaning it made the wing more airplanelike in cross section—and it increased what is called the angle of attack, the difference between the direction of the wind and the hypothetical "level line" of the animal's body. In plain words, tails on bats (just like tails on airplanes) help them fly better.

Humans care—well, *some* humans care—about the kind of tail each bat has, since different species can be identified by their characteristics, the presence or absence of a tail among them. By photographing bats when they come to feeders or taking pictures when the bats are attached to a cave roof, researchers and animal spotters alike clinch identification. Is there a long tail? The answer to that question can be the start of determining an unknown bat's species.

A Mexican free-tailed bat shows its long, flexible tail.

Hands and Discs

Many things want to eat bats, including snakes, larger bats (at least in the American tropics), and predatory birds like owls and bat falcons. At the same time, bats want to be out of the rain and be able to sleep—in a secure nest or hollow, under some loose tree bark, or in the corner of an abandoned building.

One solution is to tuck up into the underside of large leaves, which sometimes can be convinced to fold down with a few judicious bites to the frond's main ribs. For example, by cutting along the veins of heliconia leaves, Honduran white bats force

A disc-winged bat is snug as a bug in a rug, having found an unfurled leaf in which to spend the day.

the leaves to collapse into upside-down "tents" that might shelter only one bat or as many as twelve. These bats' white fur may help reflect the green color of the leaves. (It may also help them groom away ticks.)

Honduran white bats usually choose leaves for tent-making that are six feet off the ground, which is taller than a fox, coyote, or other ground predator can easily reach. These bats feast on figs, and they prefer to forage in trees that are close to their roosts. Once they find a fruiting tree, they may spend all night eating in it. In contrast, their day roosts are only used once or twice before they move on.

Another plant-roosting species, the Spix's disc-winged bat, may have a second reason to hide inside unfurled plant leaves. Disc-winged bats are named for the suction cup–like structures on their wings and feet, which allow them to cling to the smooth surfaces of their leaf roosts. A recent study finds that their roosts also serve as hearing aids, amplifying the cries of flying bats passing overhead so those inside the roost can hear them.

Noses

All bats can see and all bats can smell. Most bats also need to echolocate, either to find flying insects or to navigate in the dark, or both. At its simplest, echolocation means that a given bat is chirping out bursts of high-frequency sound and listening to the change when it bounces back. It is using three parts of its body: its mouth, nose, and ears. Compared to most kinds of diurnal animals we are used to, bats look weird. Some have radar-dish faces, some have giant ears, and some have a wedding-cake tower of folded skin rising above their mouths. Usually, beasts like tapirs that have big schnozzes are polite enough to keep their noses out front, at the ends of their faces, where we expect noses to be. Bats don't follow that rule; their embellished noses stand straight up, like a third ear or a bouquet of surprised flowers.

Common names reflect these unusual structures, and there are sword-nosed, spear-nosed, leaf-nosed, broad-nosed, long-nosed, and hairy-nosed bats. The main function of the nose "leaf" is to serve as a focusing mechanism to increase the difference between the reflected sound from a central object (like a bug or flower) and more peripheral objects.

Separate from the function of its shape, a bat's nose also does more regular, noselike things. Neotropical fruit bats are highly sensitive to fruit odors and can discriminate both odor qualities and overall quantities. Many bats use smell in combination with echolocation but appear to rely mainly on sonar cues to locate targets once their senses have been stimulated by the presence of an attractive odor cue. "The nose knows," as the expression goes.

Ears

A bat's ear shape can also act as a filter, amplifying useful echoes and diminishing the less important ones. Echolocation evolved independently in two different bat lineages, as well as in unrelated groups like rodents, cetaceans, and even a nocturnal bird, the oilbird from South America (page 90).

Echolocation is often compared to the *ping ping ping* of sonar. The actuality of it is more complicated. Bats use their mouths (oral echolocation) or noses (nasal location) to create ultrasonic sounds that bounce off of objects. This gives them information about their distance from an object and also the object's shape and composition. Those sounds are undetectable for us (or for most of us, most of the time). Humans have our best hearing from age fifteen to twenty-five, perceiving sounds in the twenty to twenty thousand hertz range. In comparison to what we can hear, bat echolocation calls usually range from fifteen kilohertz to two hundred kilohertz. So that means some humans can hear some bats some of the time, but no human ears can hear all the bats all the time.

A bat can determine how far away an object is (in part) by how fast its echo returns. These probing "feeler" calls are emitted at different intervals. A bat flying through an open field might only emit sounds once every second, and it will sync that call with its wing beat to save energy. On the other hand, a bat flying through the forest would emit calls several or even dozens of times per second, to microcorrect its flight path and avoid crashing onto trees.

How can bats navigate and hear only their own echoes in a cave when there are dozens, hundreds, or even hundreds of thousands of other bats flying at once? Two answers solve that problem. We now know that bats have incredible sensitivity to changes in the frequency of sounds. Some bats can detect changes of 0.001 kHz, and they can shift their own echolocation pitch higher or lower to differentiate their pitch from those of other bats. Bat calls are also incredibly loud—even if we cannot hear them ourselves. Bats emit sounds in the ultrasonic range up to 120 decibels, which in comparison is the same as what you would experience standing next to a jet engine. Bats avoid being deafened by their own calls by disconnecting their ear bones with the help of muscles that activate when the sounds are emitted; the bones reconnect to allow the bat to hear the echoes. This all happens multiple times per second, thousands of times in a single flight.

A bat can emit calls that jam other bats' echolocation, to beat them to a prey item. Moths, in turn, can detect approaching bats' calls and stop flying in the last moment to avoid being eaten. Sometimes they drop straight to the ground when they do this. What makes this incredible is that moths are essentially deaf. They detect these calls through the vibrations the calls produce inside their bodies, not through ears. Moths are so good at avoiding bats that certain bat species have had to change their echolocation frequencies to maintain an advantage in this biological arms race.

To glean insects that are stationary requires other tricks. Bats can flood a rainforest with sound waves and then use information from the returning echoes to navigate. Leaves reflect echolocation signals strongly, masking the weaker echoes from resting insects. So, in the thick foliage of a tropical forest, echoes from the leaves may act as a natural cloaking mechanism (known as *acoustic camouflage*) for the insects. Leaves both with and without insects strongly reflect the sound if it comes from straight ahead. When a bat approaches head-on, it cannot find

"My, what big ears you have, Grandma," said Little Red Riding Hood to the wolf. Perhaps the wolf in the bed was part bat, with a remarkable set of ears like those of this Townsend's big-eared bat.

its prey, since echoes from the leaves mask the echoes from the insect. But a recent study found that if the sound originates from oblique angles, it is reflected away from the source and the leaves act like mirrors, just as a lake reflects the surrounding forest at dusk or dawn. The angle of a bat's approach determines whether the bat can detect a resting insect.

Researchers predicted that bats would most profitably approach insects resting on leaves from angles between forty-two and seventy-eight degrees, the optimal angles for discerning whether a leaf has an insect on it or not. With high-speed cameras, scientists confirmed that this was exactly what was happening.

Size

How big are bats? Most are small or smallish (about sparrow-size), but on the other end of the spectrum, the largest flying foxes have sixteen-inch bodies and five-foot wingspans.

A reference to bees in a species's common name tends to indicate that the species is relatively small. If the smallest bird in the world is the bee hummingbird in Cuba, then the smallest bat is the bumblebee bat in Thailand and Myanmar. Another name for the bumblebee bat is "Kitti's hog-nosed bat," a name longer than the animal itself. This species is an inch long and weighs the same as two paperclips. It eats flies and spiders.

Flying foxes are found in Australia, Africa (including Madagascar), Asia, and some Pacific Islands. They are generally large, generally but not exclusively nocturnal, and they are the anomaly among bats, since most cannot echolocate. The largest bat that ever lived . . . is living now, as a matter of fact. The flying foxes of modern Earth are as big as bats have ever grown in the past. We are living in a golden age for bat species, so I hope you appreciate them.

And no, despite what is shown in horror movies, bats do not carry off small children, or at least not yet. If we keep being the nature hogs that we are, perhaps even the frugivorous bats will turn against us one day.

↘ The Indian flying fox can claim a wing-span of five feet.

▲ A great fruit-eating bat proves it can pee, fly, and probably chew gum, all at the same time.

Bathroom Break

What goes in has to come out, and in the case of fruit bats, their poo may exit in a messy slurry whose transit time, mouth to backside, took less than fifteen minutes. Bat researchers in the rainforest know that you stand under a tree full of ripe figs at your own peril. In New Guinea there is a legend that bats defecate out of their mouths, while hanging upside down. Although that idea makes an odd kind of sense, it is not true. Bats "go" the same way everything else does in the mammal world. Around colonies, though, the evidence accumulates in visible (and smell-able) ways. One easy way to check whether bats are present under free-way bridges or not is to look for the white stains of calcified urine along the edges of the structure's expansion joints.

Guano can make excellent fertilizer. But if you're in a cave and notice an accumulation of guano, you don't want to breathe in the dust, which can contain spores that cause histoplasmosis.

Sweeping out a dusty attic that has had bats warrants the same caution—open a window and wear a mask.

As far as who has what appendages, since bats are mammals, they have the same male and female parts as everybody else. If you look closely at photos, you can see that some images of bats show lactating females with pink milk ducts near their armpits, and some show males with small but distinct penises. For example, the Jamaican fruit bat pictured on page 207 is a male. Two lactating females can be seen coming to the hummingbird feeder in the pollination chapter, page 116.

Studying Bats

Researchers and community science teams can study bats several ways. Handheld detectors make bat calls audible, and you can record the files to download onto a computer later. You can also do some preliminary identification in the field. While recording a podcast in the Cosumnes River Preserve near Sacramento, California, I happened to be standing on a small bridge over a vegetated slough right after sunset. My host suggested we try the bat detector I had brought. Within ten minutes, we had heard (and dimly seen) Mexican free-tailed bats, hoary bats, Yuma myotis bats, and big brown bats. This was right after some sandhill cranes had flown over and right before we saw a western screech owl. Aside from a few mosquitoes—handled with a spritz of bug juice—it was a magical time.

There are several kinds of bat detectors. The easiest to use plugs into the charger port on a smartphone and uses a free software app to make bat calls audible and display them as a kind of bat "sonogram," with closely packed lines marking peak bursts of sound. The transcribed bat calls always remind me of the readouts from a seismograph recording a vigorous earthquake. Just as birds have different songs from each other, bats use different sound patterns and frequencies when hunting, and by noting the patterns, you can figure out what species are flying around, even if it is already fully dark out.

A similar kind of device but with larger batteries and higher capacity can be mounted on a pole or tree and will passively record bats every night for weeks or even months. Those are more of an investment to buy, but you might be able to get grants from conservation groups to defray the cost, and by placing them in parks or even backyards, you can help educate the public about the presence and value of our nocturnal neighborhood friends. In Los Angeles, urban scientist Miguel Ordeñana has been using passive bat detectors to prove that there is a diverse population of bat species all through the urban environment. Places that were long thought to be "nuked" by concrete and development turn out to be bat corridors and even major roost sites.

Bat detecting this way has been applied to bird monitoring and also to insects. Dr. Ming Kai Tan investigates the ultrasonic calls of katydids. While he was still a student at the National University of Singapore, he made good use of the COVID lockdown, employing passive monitors and collating the data they collected. He says, "During the COVID-19 pandemic, I used the opportunity to record the calls of crickets in urban areas where I was restricted to. In total, calling songs of ten species with distinct call signatures in both the time and frequency domains were recorded and were dominated by *Polionemobius taprobanensis* and *Gryllodes sigillatus*. These data allowed me to understand the acoustic community of urban orthopterans for the first time." His study was published in the journal *Bioacoustics*.

The most common way to study bats is to capture them physically in mist nets. These are like very fine-scale fishing nets (or like bird-banding nets, if you have seen those) and often are held up by poles on each side and stretched across a pond or stream. The photo on the next page shows a mist net being set up for the night at Big Thicket National Preserve, East Texas. Bats fly into the

▲ Researchers in Texas prepare a mist net for the evening, hoping to catch as many as ten species of bats.

➤ A sign at the Los Angeles Zoo offers good advice.

net, get measured and sometimes banded or fitted with tracking devices, and are then released. The bats in flight shown in this chapter had all started the evening with a few minutes' detour through a mist net. As they were released to fly out and continue foraging, they triggered a sensor linked to cameras and flash units. In essence, they took their own pictures.

Rabies

Can a bat give you rabies? Well, it *can*, but it probably won't. Less than 1 percent of all bats carry rabies, and unprovoked bat bites are rare. According to one survey, from 1995 to 2009, an average of two people per year in the United States died from contact with rabid bats. In contrast, the majority of the estimated fifty-five thousand rabies deaths worldwide each year are caused by dog bites. You can also get rabies from raccoons, skunks, foxes, and zombies. Okay, I made one of those up. But we all know a zombie when we see one, and the same rules apply for wildlife. If something is behaving strangely, or if it is (for example) a night animal out in the daytime, avoid it. Don't pick it up, don't take a selfie with it, don't try to examine it to see what might be wrong.

If handled, some bats bite. Professional wildlife biologists (including the author of this book) often get preventative rabies shots before spending time in the field, to increase their protection. In most people's day-to-day lives, though, rabies is not a problem. Enjoy animals, don't fear them.

What Bats Eat

Most bats eat insects, either caught on the wing or plucked from leaves and the ground. Some bats eat nectar and pollen, as we have seen. Some bats catch and eat frogs. Two kinds of bulldog bats in the New World tropics specialize in eating fish. Using acute hearing and pinpoint echolocation, they can detect the fish's fins breaking the surface of the water and swoop down, dragging the fish out of the water with their sharply clawed hind feet. They then fly to a perch and eat the fish while hanging head-down. A fishing bat can catch and eat as many as thirty fish in one night.

Fruit-eating bats disperse seeds across a wide range of intact and cleared rainforest, so they are important parts of any plan to conserve and regenerate degraded habitat. Anybody who says, "What good are bats?" is missing an essential truth. Bats control pests, pollinate plants, and help disperse seeds, often redepositing fruit seeds inside a gift-wrapping of "get started now" fertilizer.

Not to sound tautological, but insect-eating bats eat a *lot* of insects. Bracken Cave, in central Texas on the northern outskirts of San Antonio, is home to the world's largest bat colony. Most years there are more than twenty million Mexican free-tailed bats there. A single Mexican free-tailed bat will eat its weight in insects each night. After pups wean and start flying, the population level is at its highest. At that point, in late summer, the bats from Bracken Cave are devouring over 140 *tons* of insects every night.

▲ Bumps on the fore-head of this Jamaican fruit-eating bat look like ticks or warts but are in fact seeds from *Ficus* fruits.

◀ Pallid bats hunt ground prey, including scorpions.

My favorite bug-catching bat is the pallid bat found in the North American West. It's the official state bat of California. It is straw yellow—or as one friend calls it, dishwater blond—with tall ears and a smooshed-in pug nose. Pallid bats eat scorpions, so let's give them bravery points for that, and they also hunt crickets, centipedes, beetles, grasshoppers, cicadas, mantids, and (rarely) even lizards and mice. An injured pallid bat learned to eat meal worms in rehab, gladly taking them from the offered tweezers. I had a photo session with this bat and offered it a Jerusalem cricket (or "potato bug") from the yard, as well as a scorpion. No, nothing doing. Like a picky toddler, the bat (named Pedro by the rehab people) wanted nothing to do with those yucky things. Meal worms or nothing, that was his motto.

Vampire Bats

Some bats are obligate sanguinivores, which is a fancy way of saying they only eat blood. Or, in the case of the common vampire bat native to Central and South America, it doesn't "eat" blood so much as lap it up from a thin incision made by its razor-sharp teeth. An anti-coagulant in the bat's saliva keeps its prey's blood flowing steadily.

The vampire as monster existed before the bat; the word *vampire* (in various spellings) predates the bat's formal scientific description. Since then, each has helped keep the other current in popular culture. One thing to notice in our photo of a flying vampire bat below is how muscular its legs are. This bat's leg almost looks like the leg of a lean kangaroo. In contrast, most other bats have relatively short, weak legs. There are some exceptions, like the bulldog bats that eat fish, but most bats don't have hefty gams. The vampire bat needs strong legs because it approaches prey on the ground, walking up to a sleeping cow (for example), climbing onto its haunch, making scalpel-precise incisions, and then filling up from there. This species could, in theory, bite a sleeping person—if one were outdoors barefoot in the jungle and passed out cold, but that combination of factors is unlikely to happen very often. Historically, they fed on tapirs or peccaries, but these days they have plenty of cows and pigs to seek out. If the wound left by a bat becomes infected, that is a problem for the cow's owner (and the cow, too, of course), so many non-vampire bats are persecuted by people out of fear, ignorance, and anger.

There are two other sanguinivorous bat species besides the common vampire bat. All live in Central and South America. The

Vampire bats do exist and do lap up blood, but they are not a problem for humans. They live in the New World tropics, not Transylvania.

hairy-legged vampire bat feeds on the blood of domestic chickens and wild birds; the white-winged vampire bat mostly hunts birds but will also feed on goats, pigs, and cows, if those are more easily available. None of the three seeks out people, wears a cloak, or speaks in a fake Romanian accent.

Bat Colors

Bats are mammals, so all bats have hair and give birth to live young, and they can live for a fairly good span of time—up to ten years for a pallid bat or a wild flying fox, although twenty-year life spans are possible for the same species in captivity.

The Mauritian tomb bat can be seen on houses and trees, but rarely tombs. Apparently, nobody wanted to give it the name "hut bat."

Most bats are brown, but some are white, including the ghost bats of South America; some bats are black, while others (the "red bat" complex, for example) are cinnamon-colored. The yellow bat species complex has bats that are indeed yellow, though in Africa, one widespread fruit bat is not called yellow, but "straw-colored." The North American tricolored bat (formerly and erroneously named "eastern pipistrelle") looks light brown from far away; the three colors in the name refer to bands of color on each hair on its back. If a researcher is holding the bat for you to examine, blow gently on the bat's back to expose the colors.

Some bats' colors match their environment very well. The Mauritian tomb bat shown here is resting on the side of a hut in Madagascar. Its tan face and grizzled fur help it blend in on walls and tree trunks. Contact calls bat-to-bat during the day include chirps that are just audible to human ears. While this species is normally nocturnal, it has good eyesight, and if it decides to hunt in daylight, it is one of the few bats that also takes butterflies.

Many mammal fanciers hope to add the North American spotted bat to their life lists. It does not have small spots like those of a leopard or an Appaloosa horse, but instead sports large black-and-white squares like a flying chessboard or a pinto pony. Its giant pink ears soak up every scrap of sound; in terms of the ear-to-body-size ratio, this species has the largest ears of any North American bat. They are a rare bat—or are they? As solitary flyers that typically come out late (after midnight, some nights) and as residents of cliffs in places like the Grand Canyon, spotted bats may not be rare so much as they are rarely encountered by humans.

One guess as to the reason for their fabulous coloration is that it helps break up the shape of their body if they're roosting in a crevice that receives dappled sunlight during the day. I would prefer to think that they are this way just because it looks so good, but you are free to make up your own mind.

Bats Live in . . . Caves

Batman lives in Wayne Manor, but many other bats live in regular caves—no butlers or secret chambers involved. Caves offer a stable climate and respite from the owls and falcons that try to eat flying bats after they emerge into the upper lands. Some bats use one kind of cave to hibernate in and another for raising pups. According to the National Park Service, "Some bats roost

These European Space Agency astronauts are carrying out team-building exercises in a classic cave. The ribbons of stone are called speleothems.

alone, others in groups but not necessarily close together. Some prefer the close company of their companions and snuggle-up. Some may roost near cave entrances, while others roost thousands of feet or hundreds of meters from the exit. The bats may fly through small holes or up and down deep pits to reach their roost. Some species will roost over water such as a cave stream or lake, while others appear to always sleep up above terra firma."

If you don't get the fantods from the thought of being underground, you might want to visit Carlsbad Caverns National Park in New Mexico, a well-maintained public cave with amazing calcite features. Entering in the large natural opening, you can walk down deeper and deeper, then at the bottom, take the elevator back up. There are restrooms and a small snack bar at the bottom.

To see bats, linger on a summer evening for the ranger-guided bat emergence spectacle. No reservations are required and the program occurs every evening from Memorial Day weekend through mid-October. It takes place at the Bat Flight Amphitheater, located at the natural entrance to the main opening for

Carlsbad Cavern. The start time for the program changes as summer progresses and sunset times change, so check details when you make your entrance reservations to see the main cave.

To a bat, "caves" can include a lot of cave-adjacent habitats— what we might think of as "near-caves" or "pseudo caves." These might be naturally occurring, like slabs of rock that fall in such a way that they create talus caves, or something more human-made, like a road culvert or mineshaft. Old railroad tunnels can make good bat caves, assuming the trains now use other routes, such as in the picture above of "batters" entering a former railroad tunnel in Madagascar.

I like it when I can combine wildlife activities with cultural sightseeing. My go-to site for showing people the Eurasian house mouse is along the tracks of the London Underground, and in my last three visits to Paris and Marseilles, I have seen brown rats each time. From the cliffs of Easter Island, watch for red-tailed tropicbirds. In Cairo you can see hoopoe birds, and Singapore has monkeys, bats, and the lesser mouse-deer. For bats, the vacation to take is to the Yucatán of southeastern Mexico. Uxmal ("oosh-maul") is a city from the classical Mayan period, and it ranks as being as impressive as visiting Palenque, Chichén Itzá, or Tikal in Guatemala. Yet it has good bat sightings too, and during my visit there my guide and I found Jamaican fruit bats, black mastiff bats, and broad-eared bats.

▼ Explorers look for bats in a tunnel in Madagascar.

▲ The Maya city of Uxmal in Mexico is good for studying architecture, history, and bats (though not necessarily in that order).

Bats Live in . . . Palm Trees

Bats live in caves but also roost in trees, especially untidy ones, like palm trees with thick beards of untrimmed fronds. There are lots of reasons that people trim palm trees, from worries about liability if the serrated stems fall and hit someone to fire-risk mitigation to a basic sense that an untrimmed tree looks messy, and messiness in our culture is equated with sloth, even moral laxity. Yet palm beards serve a necessary purpose. Western yellow bats, for example, tuck up inside them to roost. Yellow bats do not mind non-native species of palms—which is good, given that most cities only offer non-natives—but their original association was with the California fan palm, which is the palm that was originally found at Palm Springs and Twentynine Palms near Joshua Tree National Park. One study by the San Diego Natural History Museum found western yellow bats in thirty-one out of thirty-two desert oases surveyed, making parks (like Anza-Borrego Desert State Park near San Diego) essential refuges for this species. A sister species, the northern yellow bat, lives in the palm trees and Spanish moss of the Texas and Florida coasts. Both kinds of bats do need to gain flight by having a bit of a drop, so very short trees, even if untrimmed, are not really suitable.

Barn owls live in palm trees like these, and sometimes great horned owls do as well.

Fan palms are native to desert oases in California, though the ones in Hollywood are introduced from Mexico. Bats roost in the fronds of both.

Bats Live in . . . Our Imaginations

If our movies and comic books are filled with drool-fanged bat-monsters and Voldemort-consorting snake-changers, that in a way could be a good thing. True, these portrayals perpetuate misconceptions and fear. They demonize things associated with nighttime and make the color black (and darkness more generally) seem bad or dangerous. On countless pulp-fiction covers, giant bats and rabid gorillas menace swooning blond women who often wear only thin negligees. It seems that one good piece of advice for would-be explorers is that if you don't want to get carried off by a giant bat, don't go hiking in a pastel nightie.

There is no question that these images have saturated our consciousness, at least in the Anglo-European tradition. Yet on the other hand, as lurid and incorrect as these portrayals are, they also remind us that imagination is a powerful force, and that if we can imagine giant devil bats, then it is equally true that we can populate our visual culture with ideas and images that present nature in general and bats in particular as something to be admired, not feared. The bats themselves are neutral; they mean whatever we make them mean. And if, in the past, we let them express our fears and resentments, then going forward it is just as probable that they can demonstrate altruism and community. Take vampire bats, for instance. They roost in social groups. The animals groom each other to strengthen social bonds, and they'll even help an ill, stuck-at-home roost-mate by regurgitating some of the blood they have just acquired. It may seem gross or strange to us to eat blood, but that's hypocritical, if so. After all, we're the ones who invented steak tartare, and we are also the ones who have the tradition that when a boy shoots his first deer, the father, uncle, or senior patriarch rubs the blood of the kill on his forehead. "There, son, now you're a real man."

Rather than fearing vampires, we should admire them. Eating blood is a brilliant ecological solution. For example, only some

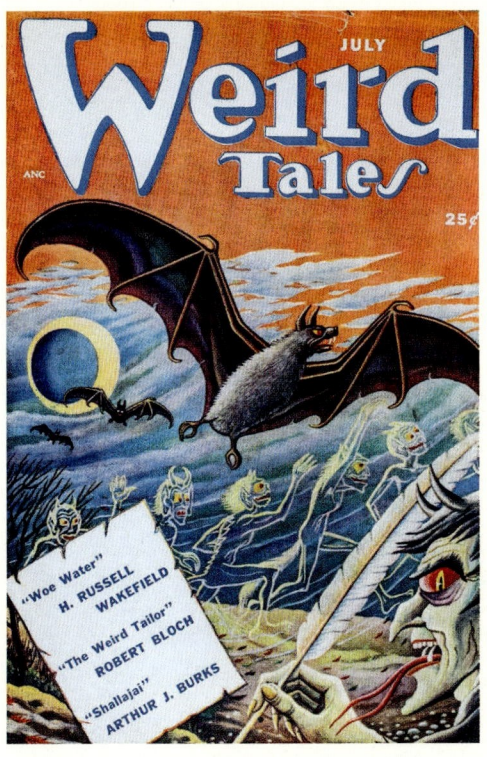

This pulp magazine cover from 1950 tells us a lot about social fears at the time and basically nothing about the biological truth of actual bats.

trees have fruit, and even that fruit is only ripe a month or two per year. Meanwhile, a sanguinivore's food supply has no limit. It is always in season and it is universally available. The only better idea would be if a bat could figure out a way to eat air itself, or how to survive by drinking only pond water. If, though, the idea of a vampire bat still creeps you out, that's fine. There are 1741 other species to think about, and surely one of them will pique your interest.

One collective noun for bats is *colony*, but I think we need to update that. To be colonized is never a good thing, and lowly bacteria come in colonies, as do ants. I propose a new term: a bouquet of bats. Or else an infinitude of bats. Or, depending on the species, a swirl, a flit, a bounce, or a bedazzlement. A pyrotechny of bats. A hustle (in the sense of a disco dance). Or, given how strange some of them look, an impossibility of bats. The Mexican free-tailed bat can accelerate to 100 mph, so we'd also have to consider a blaze of bats. If you have your detector volume switch turned too high, it would be a cacophony of bats. On radar screens, the ones leaving Bracken Cave create a blur of bats.

Whatever term we decide on, there is a bounteous bonanza of them. Go, bats, go!

Lemurs, Lorises, and Night Monkeys: Nocturnal Primates

Humans are a slim, curious leaf at the end of a long twig on the tree of life. The nearest branches include the other great apes— two species of gorilla, three species of orangutan, a bonobo, and the chimpanzee. Collectively, we eight are a group called the hominids (with a *d*); the genus *Homo*, which split off from the other apes two million years ago, is part of a bipedal sub-clan called the hominins (with an *n*).

We have much in common with our fellow apes. We all have agile brains, nimble fingers, complex social structures, a proclivity for using tools, and an unquenchable appetite for sugar and ripe fruit. Further, we all distrust the night—historically a time when there were too many leopards and not enough safe rooms—and so as the sun sets, all the other apes go up into the trees to make nests and wait for rosy-fingered dawn to come back and signal that it is safe to drop to the ground and

Panamanian night monkeys poke their heads out of a slit in a tree trunk to see what's going on.

begin foraging for bugs and sugar again. Some gorillas nest on the ground; the other species—all except humans—go up into the trees. What humans used to do to stay safe on the ground was to build a fire and sleep in staggered stages. Now, of course, we just light up the night with a million trillion electric bonfires, a holdover act of atavistic insecurity whose implications we will consider at the end of this chapter. Do we sleep better, illuminated by this planetary umbrella of nonstop light? Probably not—and the statistics about which primates sleep the best will appear later.

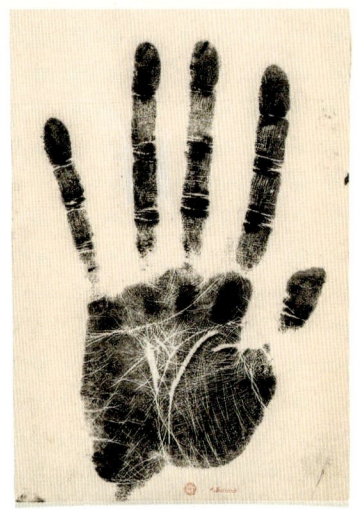

In the late 1800s, Parisian artist Henri-Charles Guérard made a print of his own left palm, titling it *Monkey's Hand*. Was he mocking Darwin or agreeing with him? Nobody is sure.

Let's start by establishing the numbers. Taxonomy for the primates is a vexed and contentious field, so the exact number of world monkeys you think are out there will depend upon which authority you listen to the most attentively. Taking a middle path between high and low claims brings us to a total of about 450 primate species alive in the world today. This number includes many of the ones we all know already, animals like baboons or howler monkeys, and also some groups that are less commonly talked about, such as the galagos and night monkeys.

As humans, we are both "of" these 450 primates and yet apart from them, and this chapter allows us to think about our relationships with the other parts of the tree of life. Our focus, though, will not be on diurnal baboons and long-armed gibbons, but on those brave and contrarian primates that primarily forage at night. Are you one of the "night owls" whose daily rhythms keep you awake when everybody else sleeps? Perhaps you will learn about the animals in this chapter and feel a bond of kinship across the eons and continents. After all, whenever we look at monkeys, whether they are diurnal or nocturnal, we see our own selfhood reflected back at us.

Nocturnal Primates in the New World

There are eleven species of night monkey in the New World, all in Central or South America, and all related to each other. They typically sleep in a hole in a tree trunk during the day, foraging at dusk until the middle of the night; then they rest, and then they forage again until sunrise. The rest of the time they sleep. For a discussion of primate hours-per-night sleeping habits, including where we fit in, see page 233.

New World night monkeys are monogamous and usually travel in small groups of several related animals. They chatter and murmur to stay in touch, but do so softly. Diurnal monkeys like howler monkeys really can hoot up a storm (and the all-black males have bright white testicles, to add to the festive display); night monkeys are quieter, shyer, and less interested in filling the air with high-octane bravado. They might sidle close to a campsite or tourist lodge to check things out, but they fade away silently if you shine a light on them or make much noise.

New World night monkeys have particularly "pop-eyed" faces, and sometimes they are called owl monkeys, since they do look like a kind of monkey-owl combo. Of course, some of that startled, bug-eyed look is a byproduct of humans shining lights on them to take pictures. What expression would you have if somebody took an industrial grade searchlight and aimed it into your bedroom window? As a corollary to that, in putting together this book, the other contributors and I often under-lit scenes or used diffused or deflected flashes to create a more painterly effect. The goal was no "deer in the headlights" kinds of pictures. If we don't want them for our own portraits, we should show the same consideration to our animal cousins.

Like most other monkeys, night monkeys eat a mix of leaves, fruit, nuts, and insects, moving through the top half of the canopy. If something is in bloom they will linger, but if not, then on they go, rarely coming down to the ground. Ocelots climb trees but jaguars don't, so staying high up means they are comparatively safer and also able to get away more quickly if they hear something coming.

What do these primates gain by switching over to the night shift? As with everything else in this book, we must assume that it's not random and that they did not wake up one day with their circadian rhythms out of whack by an even twelve hours. Predator avoidance could be part of it. During the day, harpy eagles will gladly hunt something the size of a small monkey, and while birdwatchers make a major project out of seeing this species now, in preindustrial, pre-European-arrival forests, harpy eagles were much more common. The US military airborne units have an expression, "death from above." For monkeys inhabiting forests with robust harpy eagle populations, that slogan applies with chilling accuracy. Coming out only after dark reduces that potential threat of a sudden and violent end.

Meanwhile, separate from predation, we know from the pollination chapter that some flowers only open at night, and so it makes sense if you eat flowers, the nocturnal ones will be equally desirable. For example, one food source that attracts night monkeys is the flowers of the balsa tree. These trees were originally thought to be bat-pollinated, and while they are (at least some of the time), balsa trees also are pollinated by two other nocturnal mammals, the olingo and the kinkajou. These raccoon relatives look like monkeys, and with a quick glance at night, with the animals high up in the canopy and the flashlight beam dancing back and forth, they can easily be mistaken for just that. But the olingo and the kinkajou are not closely related to primates; both are more like furry, teddy bear versions of the ringtail that was shown at the end of the cats chapter. I've been lucky enough to see olingos, kinkajous, night monkeys, and nectar bats all in the same tree at night, and while places like the Galápagos Islands or East Africa come up often on naturalists' bucket lists, the

The kinkajou looks like a night monkey and shares their love of blossoms, but it is related to raccoons.

nighttime rainforest in Nicaragua, Panama, or Peru can be just as interesting a place to visit, and often much less expensive.

A kinkajou has a prehensile tail, as do New World monkeys, some porcupines, and most opossums. Some mice and anteaters do, too. In contrast, Old World monkeys do not. This may reflect an accident of evolution, but it is more likely a response to the overall density and profile of their respective forests. The vine-rich Amazon is a good place to have a tail, while Borneo, with its more open forest interior, is a better place to be a glider, and so in Asia they have the so-called flying snakes (gliding snakes is more accurate) as well as large flying squirrels and colugos (or flying lemurs). If snakes creep you out, then the idea that snakes can fly may be even worse. Just to reassure you, I've seen flying snakes in action in Borneo, and their movement is more like a controlled bellyflop. These snakes can't (for example) maneuver

like ninja warriors and swoop into your open bedroom window. They glide to avoid predators, not to scare the tourists.

Night monkeys and kinkajous both eat insects, presumably for the protein, though by percentage, it is a less important part of their diet than fruit and nectar are. Given how many insects are nocturnal, that could be another reason for night monkeys to be out after dark. They have less competition for flowers than they would during the day, and they have better options for opportunistic bug foraging.

Nocturnal Primates in Africa and Madagascar

Some groups of animals are fairly well studied, and Jane Goodall is famous for her long-term work with chimpanzees. Other animals are more remote or more skittish or both, so species like the mandrills in West Africa are hard to study because they avoid people and live in an especially dense forest. Radio collars allow researchers to track them without needing to make constant visual contact, and they also reveal if a troop has shifted position between sunset and sunrise. Tracking on foot is hard work; it is hot and humid and the forest species of elephant—now listed as a separate kind from the "regular" savanna elephant—poses an anxiety-producing threat. Other problems occur as well. In my journal I have a note from one researcher who said in passing, "It's hard to look at mandrills quietly when army ants are biting your testicles." I can verify that this is indeed a true statement.

David Lehmann listens for signals from mandrills in Lopé National Park, in Gabon, West Africa.

African primates can be broadly separated into two groups: the mainland assemblage (chimpanzees, gorillas, baboons, vervets, and so on) and the lemurs of Madagascar. Both groups include diurnal and nocturnal species.

In mainland Africa, besides the usual daytime primates, there are five nocturnal potto species and eleven species of nocturnal galago or bush baby. This group has some confusing names. The spelling of the night bird *potoo* closely resembles the spelling of the word for the African primate called the potto. That similarity is just a coincidence, and there is also no connection between the African animal and the British lorry firm Prestons of Potto. Just to make it even more confusing, some of the pottos are called angwantibos. That label is not as commonly used as the first one, so we'll stick with potto.

More interesting than vagaries of spelling is the convergence that makes an African potto resemble an Asian loris, and an African galago look and act like an Asian tarsier. Along with lemurs, all these animals are referred to as prosimians—an earlier-evolved and less socially complex type of primate, at least when compared to the traditional apes and spider monkeys. All prosimians are nocturnal, except for some of the lemurs of Madagascar, and all are red-green color blind, as part of the enhancements that give them good night vision.

We'll start with the potto/loris group: These are more like small koalas than the conventional long-tailed, branch-swinging monkey. They have thick, woolly fur, short tails, and large eyes. The primates in this group slowly crawl around on tree branches, using both the top and bottom surfaces, as they search for insects and fruit.

All bush babies are galagos and all galagos are bush babies. They live in Africa. We've seen this pattern before: large eyes and large ears that provide great vision and sharp hearing. They can leap long distances inside the high canopy where they prefer to feed, and I've been out with expert trackers who have to run flat-out to try to keep up with them. Galagos are so adroit that they can snatch flying insects out of the air. They also eat exudate, or tree gum, of which two famous biblical kinds are frankincense and myrrh, so we can say that they have expensive tastes. Bush babies nest in tree hollows during the day and, less often, in vine tangles.

Snakes and owls hunt bush babies, and sometimes even chimpanzees do. The twenty-plus species are told apart by tail length, vocalizations, and range. Some common names for galagos—such as southern needle-clawed bush baby—can be as quirky as bird names, while others are more straightforward, as in the place-specific Zanzibar bush baby, named for the island in the Indian Ocean off the coast of Tanzania. In the Stanley

Kubrick movie *2001,* one of the lunar investigators calls home from space by using a video phone. At the time this film was released in 1968, the idea of a video phone call was impossibly futuristic. In the scene, the scientist's child (played by Kubrick's six-year-old daughter) says she wants a bush baby for her birthday. In a now-deleted scene, Dr. Floyd follows up that first video conference with a call to the pet department at Macy's to order one for her. Bush babies are now endangered, plus they are only active at night, so file that scene under the heading, "Don't try this at home."

Lemurs only occur in Madagascar, as one example of the island's rich offering of endemic plants, reptiles, birds, and mammals. Even well-traveled birders add big numbers to their lists by visiting, since 44 percent of Madagascar's resident birds are endemic.

There are no conventional kinds of monkeys native to Madagascar, but there are a lot of lemurs. According to one review of the region's biogeography, "phylogenetic, genetic, and anatomical evidence all suggest that lemurs split from other primates on Africa around sixty-two million years ago and that the ancestral lemur lineage had dispersed to Madagascar by around fifty-four million years ago. Once on the island, the lemur lineage diversified. Now there are at least fifty species." That was written in 2009, and since then, the number of recognized lemur species has doubled; most taxonomies now push the total lemur number past one hundred.

This brash bush baby is raiding a game lodge's dinner table, but usually they eat sap and insects in the tops of trees.

About half of the lemurs are diurnal and half nocturnal. The daytime ones get most of the attention; the Verraux's sifaka shown in the photo here populates hundreds of YouTube clips because it hops upright when moving between trees. The gait is a sort of sideways lope, hands held high, like a fussy C-3PO trying to cross scalding mud. Another well-known export from Madagascar, ring-tailed lemurs are social and diurnal, making them good choices

A Verreaux's sifaka catches the last of the afternoon light while lounging on a vine in Madagascar.

for zoos. These are the gray kind with raccoon masks and white faces, yellow eyes, and long, thick, starkly black-and-white tails. They walk like coatis (in case you've seen a coati and can make the comparison), using all four legs and keeping their tails stiff and tall. Many people know ring-tailed lemurs from cinema (the *Madagascar* franchise) or from television, since there must be about a thousand nature documentaries featuring them. As the joke goes, if lemurs didn't already exist, the BBC would have had to invent them.

The smaller, more nocturnal lemurs can be just as fascinating. Does the extra effort needed to seek out a rare or inaccessible species make it even more joyous when the search is successful? Jon Hall, the globetrotting, binocular-toting king of all the mammal spotters, recently went back to Madagascar to try to see the aye-aye, an especially odd lemur. This is the one that you have seen online probably, with black bat ears and a long, skeletal middle finger that is used to wiggle insects out of holes. Duke University's Lemur Center calls it "the strangest primate in the world." If threatened, aye-ayes can raise their white guard hairs up to look fierce. Jon Hall has seen more than two thousand mammals in the wild, and while he would say that all animals matter equally, he also would agree with Orwell that some animals are more equal than others. For him, seeing this lemur species was part of a lifelong quest. Jon shared his impressions with me after he was finally lucky enough to find one: "The aye-aye is uniquely bizarre: part Dobby the house elf, part drag queen. And my thrill of seeing one in the wild is as difficult to describe as the creature itself."

I have had my own lemur quest. For me it was not a particular species that was the goal, but a specific number. Fewer than a dozen people worldwide have seen more than one thousand species of mammals in the wild. (Jon Hall's list is so far past the rest of ours, it doesn't come up when we "runners up" compare aspirations.) As I edged closer to the goal, I wondered how I could add the final few sightings to my list. I knew that if I went to Madagascar, there would be enough lemurs for me to see that I would surely make it past the finish line. I even made up a sign ahead of time with "1,000" written on it, to hold up at the moment of triumph.

It took a few nights, but finally, there it was: Crossley's dwarf lemur, my number one thousand, posing in the spotlight, the most beautiful animal in the world. I was shaking so much I could barely hold my light. Taking the sign out of my daypack, I asked my friend José Gabriel to take a photo of the beast right away, before it could scamper off, and then to take a photo of me.

◄ The author pauses on a midnight hike in Madagascar to celebrate seeing his one-thousandth mammal.

►► Crossley's dwarf lemur is a nocturnal primate from Madagascar, and by the luck of the draw, this one was the author's one thousandth world mammal.

I am always interested in how pieces of nature get their names. Alfred Crossley collected animals for museums in the 1860s; besides gracing this lemur, his name lives on in the names of butterflies, moths, and three species of birds. Some of the night lemurs hibernate during the dry season, not so much avoiding the cold as waiting until the rains come back and help the forests fill with food again. Crossley's is one of the dry-season hibernators, one reason why so many of the nature tours schedule their travel seasons in October and November.

Have you ever thought about why kittens and puppies are so cute? It has something to do with oversize paws, giant "pet me" eyes, and a babylike, smooshed-in face. Plush fur helps, too; it's one reason a panda is considered cuter than an iguana. The large, round eyes suck you in: You must be stout-hearted indeed to be able to attend a pet adoption event and not come out with two or three new family members. That said, in a survey of cute animals, the cutest of all the cute animals in this book has to be the gray mouse lemur.

To be a mouse lemur is to be one of twenty-four species; when a lot of species look alike but are shown through DNA and other evidence to be different from each other, sometimes they are said to be "cryptic" species. That means they may look so identical that they have been lumped together for years as the same species, with the different variations all "hiding in plain sight." Vocalizations help differentiate them, as does dentition. Would the collective title for these twenty-plus kinds of mouse lemurs be "mice lemurs"? No matter what we call them, an individual mouse lemur would fit in your cupped hands. Some weigh only two ounces. All together, these species represent the smallest primates in the world.

Eating leaves is called being folivorous, and it's a great food strategy . . . and yet also a terrible one. While there are a lot of leaves in the typical forest, only a few are the tenderest, moistest, most succulent ones. Mature leaves are full of cellulose and

"Very cute and very small"— two ways to describe the gray mouse lemur of Madagascar. It would fit inside a large coffee cup.

toxins, and they can be leathery or even completely dry. Because mature leaves offer so little return on energy invested, animals that have evolved to eat leaves need to be able to slow their metabolisms. That is in part because it takes a long time to digest this kind of food and in part just to conserve energy in the first place. Koalas and sloths come to mind as animals that would be good at this, but some lemurs do it, too, trading off a slow-moving lifestyle for the reliability of the forest's greatest export.

The species shown on page 231, the red-tailed sportive lemur—which does have a vaguely reddish tail—has the slowest known metabolism of any known mammal. It can drop into a torpid state when not alert and not feeding—not quite full hibernation, yet not quite full wakefulness either. This allows them to conserve energy and take the time needed to process their tough, fibrous food choice. The opposite approach would be insect-eating shrews. Some shrews have hearts that beat 1200 times a minute; their total lifespan is one year long. "Live fast, die young," seems to the shrew's motto. This lemur would ask, "What's the hurry? Just be chill."

Sportive lemurs are midsize nocturnal primates weighing a few pounds and, in the case of the species shown on page 231, tending to forage vertically. We don't often use the word *sportive* anymore, but it means lively, playful, agile—similar to but not an exact match for what today we call "sporty." Many animal names preserve lost idioms, as this lemur's does or as in the name of the North American warbler called a blackpoll. We think of a poll as a survey conducted over the phone, but it used to mean "cap" or "top of the head." One kind of African antelope, the bay duiker, is named not for a body of water but for a red-brown horse. The gray American shorebird with black-and-white wings, the willet, was once called by the Puritans a humility, since it only shows its finest plumage to God. The usual name, willet, comes from its startled call. As the poet Mary Oliver says, "attention without feeling is only a report."

Nocturnal Primates in Asia

Africa has pottos; Asia has lorises, a near equivalent. As we were going to press some updates came in, so here is the current scorecard: pygmy lorises (two), slender lorises (three), and slow lorises (ten). All move slowly and deliberately, making almost no noise, hence the term *slow* loris. If threatened, they stop and remain motionless. A loris has arms and legs of the same length, as well as a flexible spine—at times watching them it seems that they are like furry Slinkies, but a Slinky that is stuck in molasses and only can do one "hop" per minute.

Individuals communicate by whistling and by scent marking. Lorises don't like to be out in moonlight, as I experienced once in Sri Lanka, when my visit coincided with a full moon. My guide Uditha took me to his favorite, "sure bet" loris forest, only it was empty—the lorises were all inside their tree holes, waiting for the moon to set. He told me that in twelve visits, he had never *not* seen lorises there. It also may have been too chilly, since these creatures prefer warm nights.

Lorises eat insects and tree gum in about equal measure, and they will use their claws to "mine" a tree and make it produce sap from the incised scores. No maple trees grow natively in their Asian forests. If they were to taste maple sap for the first time, would they be giddy with delight? Maybe they would find it too sweet, or maybe not sweet enough—how animals perceive the world is mostly a mystery to us.

Here's a question even Wikipedia can't answer: Are all lorises venomous, or just one (or both) of the pygmy kinds? Sources disagree. According to the San Diego Zoo, the pygmy slow loris "is the only known venomous primate, a highly unusual characteristic among mammals. The loris produces a secretion from glands on the insides of its elbows, which, when mixed with its saliva, serves to venomize its bite. Female lorises sometimes 'park' their youngsters while they feed, so they whip up the toxic elixir and lick the little one's fur to deter predators." It could be that all loris species have similar traits and we just have not yet checked thoroughly. Other venomous mammals besides the loris group include the duck-billed platypus, some shrews and moles, and solenodons, which are shrewlike mammals in the Caribbean, related to hedgehogs.

After the lorises, maritime Asia's other nocturnal primate group is the tarsiers, spread out over fourteen species. That number may rise as more detailed studies are carried out.

Tarsiers are baseball-size, with immense eyes. In some species, the eyes are bigger than their brains. They have the largest eye-to-face ratio of any living mammal. And like the galagos of Africa, tarsiers have spring-loaded back legs and can jump fast and far. By far, we mean *far*: more than forty times their body length. And no tree sap for them: These are super-swift insect snatchers that hunt beetles, spiders, walking sticks, and grasshoppers. If they can catch a bat or lizard, they will eat that, too. They live and forage the lower, denser part of the forest, between three and six feet off the ground. They want to avoid eagles and owls and yet have access to as much food as possible.

These are not animals you can see in zoos, so to experience a tarsier means going out at night in the forest. So far as I know, no one has seen every tarsier species in the wild, so there is a challenge waiting to be completed by the right person. Perhaps it will be you, and if so, please write a book about your adventures. I, for one, want to hear about them.

Are tarsiers "us" in any kind of way? We humans have faith traditions that provide us with a wide variety of origin stories. According to Juan Lozada, chief of the Kitanemuk tribe in Southern California, "We were here on the day the first sun came up." To me, science and faith do not exclude each other, and we can have absolute divinity and ancient DNA equally present in the same bundle of hope and possibility. Evolutionists would say that when we study the tarsiers, we are looking at some of the earliest rough drafts of *Homo sapiens*: Their genes and ours overlap, and the tarsiers extend back to fossils that are fifty-five million years old. We even talk the same. Tarsiers sing to each other! According to Dena Jane Clink, using computer analysis of their calls from Sulawesi, researchers "could tell individual tarsiers apart based on their songs. Being able to recognize who is singing from far away may be an important function of tarsier songs. We also found that if a female speeds up her song, then the male speeds up his song, too. The ability to modify vocal output based on what others are doing is a universal in human language. Our results show that tarsiers (like humans) can change their vocalizations based on what their partner is doing."

That all sounds like a sisterhood of fellow travelers to me. Dr. Clink adds, "The fact that tarsiers and humans are both able to do this indicates that their common ancestor probably had this ability. Our results add support to the idea that flexibility in vocal interactions evolved long before the appearance of modern humans."

Do the Other Monkeys Sleep the Same Way We Do?

The short answer is "no." All the other primates seem to sleep more (and perhaps better) than we humans do. According to a chart published in the *American Journal of Physical Anthropology*, primate sleep spans range from almost seventeen hours out of every twenty-four hours for the three-striped night monkey to eleven for a ring-tailed lemur, on down to nine and a half for a chimpanzee. In this study, only humans came in at fewer than eight hours per night.

Yet a medium-long answer might be that we are not so bad at sleeping after all. It could be that we spend fewer hours asleep than our nearest relatives, and yet more of our night in the phase of sleep known as rapid eye movement, or REM. Even in preindustrial societies—at least in the ones studied recently—people do not sleep ten hours a night, and instead the average is between five and seven hours. How did humans manage to compress the amount of shut-eye needed, compared to other primates?

Answer: We sleep less in total quantity, but we spend more time (proportionally) in REM sleep. Early hominins sleeping on the ground by a fire could stay awake longer—which kept them more secure from predators—and they also could watch out for each other, with some awake and some asleep. It's possible that those who stayed awake could make up a sleep deficit with diurnal naps. This method of getting out of the trees and into a ground-based community offers multiple benefits. Talking around the fire allows one to share wisdom and process the day's events, for one. That would make these hominins more successful, ecologically speaking. And so, with the rise of society comes a rise in communal support and the ability to inhabit riskier habitats, such as open savanna.

Don't sleep more, just sleep better—that may have been the original hominin strategy.

Is Too Much Light Ruining Our Sleep?

We've been so successful at building electric bonfires along every street, backyard, and motorway; around every used car lot and junkyard; and across all the baseball diamonds, landfills, and water treatment plants, from coast to coast and nation to nation, that now the entire world is ablaze with light—all night and every night. In some cities it is a struggle even to see Venus and three or four stars. While our human ancestors twisted sticks and used minute piles of tinder to start modest fires for cooking food and warning the big cats to stay away, we now gorge ourselves on more light in one night than previous generations saw over the course of an entire human lifetime. This has consequences. Migrating birds fly into overly lit buildings while our power plants run at maximum capacity seven days a week trying to meet our voracious appetite for *juice*, more and more juice. We love electricity even more than we love sugar.

Do we really need all that light? Most malls close at 9:00 p.m., most baseball games are over by 10:00 p.m., and most freeway traffic eases up by midnight. What if we started to turn the lights off as we left the room? Not just in our houses, but for instance at a mall: Do we need to light empty asphalt? (It's not as if the parking spaces care whether they are lit or not.) What if there were only half as many lights on in the typical mall parking lot between, say, two and six in the morning? Or in another example, I've seen cars on the freeway at night with their headlights off—not because the drivers were drunk, but because freeways are so brightly lit it seemed to them that their car lights were on already.

We have left some of our primate heritage behind and can be bolder now. Tarsiers are cute, certainly, but it is also true that they are very small in relation to all the things in the world that might want to try to eat them. They do have a reason to be cautious at night generally, and especially to be wary of being caught out away from the trees on a moonlit night.

And then there is us. We are tall, strong, clever bipeds with few extant natural enemies. We drive two-thousand-pound cars armed with Xenon headlights that are as bright as small galaxies. In some US states, we are armed to the gills with semiautomatic weaponry. Yet we have been trained to think there is no such thing as enough light. More streetlights! Higher high-beams! The reality is, we have way more light than we need. ("Surplus to requirements," as they say in England.) The same "eco" family who would never leave a garden hose running in the driveway for ten hours straight will gladly leave the porch light on all night. In squandering so much light needlessly, we are ruining night skies around the world, wasting expensive energy, and cutting ourselves off from our natural heritage.

This photograph of Spain and Portugal taken from the International Space Station shows a well-lit Europe. Gibraltar is lower right (blurred by some ground fog); ships at sea show up as white dots. It is beautiful, but do we need every single one of those lights on in Lisbon and Madrid? Probably not—the airline pilots and UFO saucer drivers all can find their way home without them.

The International Space Station passes over Portugal, Spain, and Gibraltar on a calm but overlit night.

Dolphins, Diatoms, and the Great Elevator of Life: Oceans after Dark

I n 1871, the impresario P. T. Barnum took over a struggling circus and began to advertise it as "The Greatest Show On Earth." For his guests who had never seen tigers, lions, bears, or elephants—not to mention trapeze artists in flashy sequins—that bombastic claim was probably true: It *was* (for them) the greatest show they had ever seen.

No disrespect to Barnum, Bailey, or the Ringling brothers, but I would say that the greatest show on earth is the Earth itself, with all its drama and splendor, from spectacular sunsets to supercell thunderstorms to the minor miracles of glow-in-the-dark mushrooms and cascading meteor showers.

And here on Earth today, the greatest show happening now—and one that happens every single night—is mostly hidden from us because it takes place inside the ocean.

A half-inch-long gossamer worm is captured by a remote camera, nine hundred feet below the surface. When disturbed, it can release bioluminescent particles to distract predators.

The spectacle starts with the basic reality of photosynthesis: Life is based on light. Phytoplankton, which are also known as microalgae, are the basis of the marine food chain. The term gathers up thousands and thousands of species into one plural noun. This is small-scale ecology; some of the constituent organisms are only 1/25,000th of an inch long.

Phytoplankton are similar to land plants in that they contain chlorophyll and they use energy from the sun to combine carbon dioxide and nutrients into carbohydrates. Everything else depends on them. As biologist Philip Mladenov explains, "Roughly half of the planet's primary production—the synthesis of organic matter by chlorophyll-bearing organisms using light energy from the sun—is produced within the Global Ocean." He adds, "These energy-fixing microorganisms—the marine environment's invisible pasture—form the basis of the marine food web, the network of pathways through which food energy is transferred to all the other organisms in the marine system including other microbes, zooplankton, fish, marine mammals, and, ultimately, humans." The scale of this is almost beyond our usual units of measure. According to the MIT Climate Portal, "There are a billion billion billion phytoplankton in the world's oceans." Further, "phytoplankton are hugely diverse, with likely one hundred thousand different species." This happens all around the globe. As sunset and sunrise slide from east to west every twenty-four hours—across the Pacific Ocean, then the Indian, the Southern, and the Atlantic—swarm after swarm make the same upward journey, retreating as daylight returns.

The food chain connects with the carbon cycle, as further described here by the MIT Climate Portal. "As phytoplankton die, and the creatures above them in the food web die or poop, a small fraction of the carbon the phytoplankton took in during their lifetimes sinks down below the [sunlit] layers of the ocean. As this organic matter sinks, it becomes food for deeper dwelling animals, but also nourishment for bacteria. ... This slow movement of carbon—from the atmosphere, through phytoplankton, up the food chain, and back into the deep ocean—is called the *biological pump*, and is a key player in the Earth's carbon cycle."

The problem with this inverted pyramid of planktonic wonderfulness is that all the light is available only at the ocean's surface. That is because within the first thirty feet, water absorbs more than 50 percent of visible light. Even in clear tropical water, only about 1 percent of visible light—mostly in the blue range—penetrates down to three hundred feet. The ocean itself averages twelve thousand feet deep, and it sinks much deeper than that in some sections. So that means photosynthesis is limited to just the thinnest slice of the uppermost layers. (If we think of the ocean in cross section, the sunlit layer, in relation to the unlit part of the sea, is the same ratio as the skin of a grape is to the rest of the grape, only even thinner.) For the photosynthesizers themselves, that's fine. But for the things that want to ingest them, that creates a problem, since once you rise up into the clear, well-lit top layer, you're suddenly visible to terns, sharks, whales, and fishermen. You want to eat phytoplankton, but no matter how small or transparent you are, you're in the danger zone up there, since everything else wants to eat *you* in return. Safer (much safer) to spend the day hiding out in the dim, cool, lower levels—at least to stay there until it is safe to move about. That happens after dark.

And so begins the great elevator of life. Every night trillions of animals (mostly smaller-than-a-rice-grain zooplankton, but other, larger things too) surge upward from the middle layers of the ocean to feed at or very near the nutrient-rich, plankton-dense surface. Some of them cover distances as great as three thousand feet to reach the surface. They stay there all night, and then just before dawn, they sink, swim, pulse, or glide back down to the colder, darker, less productive (but still safer) middle layers of the ocean, many hundreds of feet below. Smaller fish and squid follow the zooplankton up and back down, and other bigger fish follow the lesser fish, with the predators tracking prey the way lions stalk wildebeests. This vast oceanic migration happens out of sight and hence (mostly) out of mind; few people appreciate the scale of the phenomenon and the role it plays in maintaining planetary health. At the same time when the hunters are rising and falling, some of the phytoplankton are making opposite movements; at night, they try to sink lower in the water column. (For them, there are often more nutrients lower down than at the surface.) The nightly phenomenon is known in ocean ecology as "diel vertical migration," or DVM. Somebody please ring the marketing department, because anything this grand and important deserves a better acronym.

While this starts with microscopic organisms, animals of every size are involved. Many dolphins hunt at night, for example. We don't think about that much, since most of us only see dolphins in daylight, either from whale-watching boats or at marine parks, or less commonly, right from shore. But if the squid arrive at night, then night is when the dolphins need to hunt. The mother and calf Atlantic spotted dolphins shown on the next page are part of a larger nocturnal hunting party. Their pod-mates have helped round up a ball of fish, and the dolphins will take turns making strafing runs through it while the others "ride herd." Steve N. G. Howell, a bird and flying fish expert, talks about being in a rubber Zodiac boat near Ascension Island in the

Dolphins are among the predators patrolling the DVM. They hunt coop- eratively at night and then rest, play, and bow-ride during the day.

mid-Atlantic. He and the other passengers were going ashore to see turtles lay their eggs. It was still dark, and the local bottle- nose dolphins were hunting. "The elegant, playful, bow-riding dolphins of the daytime were changed ... into lightning-fast killing machines that slashed the water into phosphorescent streaks," he says. To escape, the flying fish would burst out of the water like rockets and "fly" away—really, just an accelerated glide—and in their random bursts, a few even landed in the tourists' boat.

With DVM, so many animals are moving at once that it actually stirs the water like a kind of giant, slow-motion mixing spoon. On early radar scans when oceanographers were trying to map the sea floor, the fish and zooplankton resting midday were so densely layered that they showed up as solid strata, as if the soundwaves were pinging off of underwater summits and mesas. But it was, of course, not new topography after all— Atlantis would have to wait for another day to write itself onto the map. The researchers had discovered an immense layer of biomass waiting for the sun to go down so they all could com- mute to work.

How Does the Ocean Work?

Oceanographers think of ocean structure in two ways: vertically, so that you can label the layers according to depth and salinity (and usually temperature), and in terms of horizontal flow, as the water from one section flows into another. According to ocean-ographer Dorrik Stow, "Ocean currents are many thousand times more powerful than any river on land. Those in the deep ocean are silent and unseen. Here, there are waterfalls without sound, rivers without banks, and storms that rage unnoticed for weeks at a time."

We mostly notice only the surface effects, and usually only those that reach land: "surf's up!" as they used to say in Malibu. (Now when they say it, they're worried about coastal erosion undermining their homes.) The great break that somebody catches at the Trestles in San Diego or on the Dare County beaches of the Outer Banks of North Carolina originated far from land and crossed a lot of the map to turn into a surfable break. "Major storm waves generated in the Southern Ocean off Antarc-tica can take nearly a week crossing the Pacific before they break along the shores of Hawaii," explains Stow. "Those that miss the islands travel on for another three or four days before they wash ashore on the remote beaches of Alaska."

The seas are never static. Currents flow and change; waves cross long distances; blobs of warm water are displaced by colder ones. Even salinity changes from one part of the planet to another. The Mediterranean has high salinity because there is more evaporation than there is rain or extra fresh water added from rivers. The seawater adjacent to Antarctica has lower salin-ity, since that area has thawing glaciers but no rivers with thick sediment to add salts back in.

The chemistry involved creates a complicated dance. The density of seawater increases with increasing salinity, and the salt content of seawater simultaneously alters the relationship between temperature and density. Salt also depresses the freezing point of seawater and may inhibit sea ice formation in salty oceans, so if water composition changes significantly, that in turn will influence sea ice more broadly.

Using temperature, pressure, and light, oceanographers divide the cross section of a given slice of ocean into named layers. The epipelagic zone extends from the surface to 660 feet, or 200 meters. It is in this zone that most of the visible light exists. With that light also comes heat from the sun, which is responsible for wide variations in temperature across this zone, both with the seasons and with latitudes. Sea surface temperatures range from the nineties Fahrenheit in the Persian Gulf to twenty-eight degrees near the North Pole. Winds and currents help mix the water, but this region has the greatest variation of temperature from top to bottom.

The mesopelagic zone goes from the bottom of the first layer down to about 3300 feet (1000 meters—science likes to use even numbers). It is sometimes referred to as the twilight zone or the midwater zone, as sunlight this deep is very faint. Temperature changes are the greatest in this zone because it contains the thermocline, a region where water temperature decreases rapidly with increasing depth, forming a transition layer between the mixed layer at the surface and deeper water. Because of the lack of light, bioluminescence begins to appear on organisms in this zone. The eyes on the fishes are also larger and generally upward directed, the better to see silhouettes of other animals against the dim light.

The depths from 3300 to 13,000 feet comprise the bathypelagic zone. The wreck of the *Titanic* rests at this depth. Due to its constant darkness, this zone is also called the midnight zone. The only light at this depth and lower comes from the bioluminescence of the animals themselves. The temperature here is constant—a chilling thirty-nine degrees. The pressure in the bathypelagic zone is extreme: over 5850 pounds per square inch.

Yet some surface animals can survive it; both sperm whales and Cuvier's beaked whales can dive this deep. How they navigate and avoid getting the bends (a fatal build-up of nitrogen in the blood) remains a topic of much interest in the marine mammal community. For whales, it seems their lungs compress in ways that ours cannot; they basically manage gas flow by limiting the "full" part of their lungs.

Past even the midnight zone, the abyssal zone extends down to twenty thousand feet below the surface. In comparison, the tallest mountain in North America, Denali in Alaska, is 20,310 feet tall. Now we have reached the pitch-black bottom layer of the ocean. The water temperature is constantly near freezing, and only a few creatures can be found at these crushing depths. And yet we are still not at the bottom-most part of the bottom. The deepest of the deep is the hadal zone, bottoming out at 36,198 feet in the Mariana Trench off the coast of Japan. The weight of all the water overhead in the Mariana Trench is greater than eight tons per square inch.

Even here, life exists. In 2005, tiny single-celled organisms called foraminifera, a type of plankton, were discovered in the Challenger Deep trench southwest of Guam in the Pacific Ocean. The deepest fish have ever been found was in the Puerto Rico Trench at 27,460 feet. The usual dismissive comment is that we know more about the moon than we do the bottom of the sea. That is true only for cartography. The surface of the moon has been mapped in very fine detail, and the seafloor overall has not (yet). But in terms of the total quantity of archived and documented knowledge, we do know a heck of a lot about water. As the oceanographer Helen Czerski reminds us, "We absolutely know more about the deep ocean [than the moon] because there is *more to know*"—the emphasis is hers.

The discrepancy is not in the knowledge itself, but in the gap between what we know so far versus what we still want to know. I will invent a hypothetical unit of measure, the gunnysack. For the moon, we have one hundred gunnysacks of knowledge so far. And then for the ocean, we have one thousand gunnysacks. And for where we want to be in the future, with the moon, we only have a few hundred gunnysacks left to go. And for the ocean, while we do have the one thousand gunnysacks so far, we have many thousands of gunnysacks of questions still to answer. That doesn't negate the first thousand gunnysacks. They are still full and extant. And as science fills in knowledge more completely, as a byproduct of that, I am sure the seafloor maps will catch up to the lunar maps, and in the end, will even surpass them.

Where Did the Ocean Come From?

We know that Mars, Venus, and the moon do not have liquid water oceans on them, and we also know that volcanoes on Earth spew out liquid rock, not liquid water. So, leaving the accounts of Genesis out of it just for now, are there any other ideas about where all this water came from?

And just to be clear, there is a *lot* of water on our planet. Separate from lakes and rivers and swimming pools, when it comes to oceans, the National Oceanic and Atmospheric Administration estimates that the Earth holds 321,003,271 cubic miles of seawater. That may be hard to picture, but a cubic mile is literally a giant square of solid water, one mile by one mile by one mile. Now multiply that by 300 million, and becomes clear that we are a very, very water-rich planet, at least in terms of seawater. It may not always go where we want it to go, but calm or wavy, high tide or low, there is a heck of a lot of it.

The salt in seawater comes from the Earth's own geology, dissolved and spread by erosion and rivers, but where the water came from in the first place takes us back to an earlier and more dynamic solar system. Right now, things are stable, other than I am sorry poor Pluto got booted from the Fellowship of the Nine

Planets. Earlier, though—as in, four billion years ago earlier—our planet endured what even sober-minded scientists refer to as the "Late Heavy Bombardment." Early in Earth's history, a lot of space stuff crashed into other space stuff, and that stuff bashed into Earth, and so we were hit by unending salvos of comets, asteroids, failed planets, planetesimals, and debris. Water came to the planet via ice on comets and asteroids. In this narrative, most or all of our seawater is extraterrestrial.

That is a great story. Too bad that it may be a load of hooey.

Newer ideas suggest that the bombardment phase was not that prolonged or intense, and that our planet's water could have been generated by interactions between the hydrogen-rich atmosphere and magma oceans that covered the surface during Earth's formative years. An article posted by Carnegie Science presents evidence that "early in Earth's existence, interactions between the magma ocean and a molecular hydrogen proto-atmosphere could have given rise to some of Earth's signature features, such as its abundance of water and its overall oxidized state." Computer modeling and studies of exoplanets have backed up the idea that interactions between magma and

Did all the water on Earth arrive via comets crashing into us? Or was it created onsite, the result of chemical reactions with hydrogen? The first possibility creates better graphics, such as this one from NASA.

the atmosphere resulted in the movement of large masses of hydrogen into the metallic core, the oxidation of the mantle, and the production of large quantities of water.

I will be a bit sad if this turns out to be the correct answer. (And we should add quickly that there is no reason to think it won't be.) It is just that I will miss the old illustrations, with their emphasis on crashing fireballs and geysers of molten rock. What may be the right answer just also happens to be the most sedate answer.

How Do the Animals Know What Time It Is?

If you've ever had pet fish, you will have noticed that they can tell when it's feeding time. Partly that is external clues—your alarm has just gone off, the room lights have come on, and everybody in the house is scurrying about doing the morning hustle—and partly it is from internal signals, since all (or almost all) living beings possess internal clocks.

For DVM to happen at roughly the same time each day, all the fish, squid, and zooplankton at lower depths need to know when it's time to start heading upstairs. Circadian rhythms not only govern the awake/sleepy divide, but also control daily rhythms in body temperature and in our cortisol and melatonin production. In fancy words, "Circadian rhythms are based on endogenous clocks capable of generating oscillations and of imposing this rhythmicity on downstream entities"—that is from a 2013 study.

There is of course plasticity in this mechanism. The arctic terns that we mentioned in the migration chapter need to be able to forage constantly in the continuous daylight of the Arctic summer, and then they need to rest normally as they transition through the equatorial regions, which have fairly consistent twelve-hour days. Some bees seem to have no circadian rhythm.

Honeybees that are active around the clock have intermittent sleep that is distributed throughout the day in association with times of inactivity. Blind cave fish seem to have lost their circadian rhythm patterns, and some open-ocean sharks seem to need less sleep than typical ones. Meanwhile, most circadian animals can adapt to temporary disruptions in the expected pattern. Mama whales need to help young calves breathe, so even more than human mothers, they are tasked with around-the-clock care for the initial few weeks. Like migrating birds, they seem to be able to adjust to the lessened rest without showing signs of extreme stress or fatigue. Those exceptions aside, in most mammals there seems to be a master clock in the brain that syncs with neuronal, endocrinal, and paracrinal signals, and all of those pacesetters cross-calibrate with external cues, such as the quality of daylight coming through the blinds as you're starting to wake up. It's a centralized system that is connected into a feedback loop with additional sensory data.

Even a partial eclipse of the sun (when the sun is still much too bright to look at directly) alters the human perception of daylight.

As humans, we analyze these clues and participate in micro-evaluations without even thinking about it. On the day when I was working on this chapter, there was a partial eclipse of the sun, and seeing me standing on the sidewalk looking up, one of my neighbors came out to ask me what was happening. She had not heard anything on the news but could tell there was something "off" or "odd" about the light—the day felt wrong to her. Even the shadows from the leaves on my desert willow tree were not quite right. I told her there was a partial eclipse going on and loaned her my safety glass so she could see it herself. "Ah," she said. "That explains it!"

A crystal jelly heading to the surface looks like a lonely spaceship crossing the infinite void between one star and another.

Evidence of the Great Migration

The DVM is large and constant, but how can we see it for ourselves? The oceans are large, the night is dark, and the participants are, on average, microscopic. How do we know it is going on? One way to is to look for secondary evidence the next day.

When the DVM happens, everything is in motion. As the sun goes down, the sharks come up. There are multiple shark species, of course, but we are interested in one particular kind: cookiecutter sharks. Their name comes from the circle of teeth lining their mouths. Instead of cutting a star or Christmas tree shape out of a sheet of sugar dough, these two-foot-long sharks cut a circle out of their victims that is four inches across and

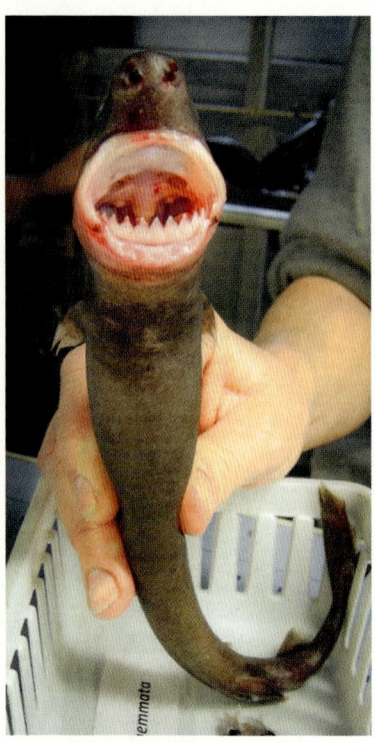

several inches deep. Once you know to look for evidence of their attacks, you will see it on many animals. In the photo above of a rough-toothed dolphin, the white, flower-shaped circles on the flanks are scars from cookiecutter bites. As a potential prey item swims past, these parasitic sharks dart in, latch on with sucker-grip lips, and spin in a circle to rip out a biscuit of flesh. They peel away before the dolphins can counterattack. It's a simple attack plan: bite, twist, flee. Repeat as needed. A search online will turn up images of seals, whales, and dolphins, all with the round, jagged scars left over from these smash-and-grab attacks.

For their victims, the individual wound site may not be huge or deep, but if it gets infected, the animal will be weakened, and in the case of the dolphin shown above, it obviously has been bitten multiple times in a row. As a side note, the pink color on the tummy is a network of thin blood vessels as the well-insulated

⮊ A rough-toothed dolphin breaches in Indonesia, revealing a row of white, flower-shaped scars from cookiecutter shark bites.

▲ This cookiecutter shark looks like an alien from a scary sci-fi movie.

In the 1970s, American submarines were damaged when cookiecutter sharks took bites out of their accessories' neoprene housings.

dolphin "blushes" to cool off. The "rough-toothed" part of the dolphin's name is only evident in a deceased one on a dissecting table; the teeth have grooves not present in the mouths of other species. The scratches on the head of this individual are probably from tussling with other rough-toothed dolphins, though some could be scars from squid beaks, as dinner fought back against diner in an MMA battle of cetacean versus cephalopod.

Of course, compared to other targets, whales are small potatoes. Forget whales and dolphins—in an ultimate case of top bragging rights, cookiecutter sharks are famous for disabling American submarines. During the Cold War, the largest submarines in the US fleet belonged to the Ohio class of ballistic missile submarines, including the *USS Kentucky*, shown above docking in Guam. These submarines were among the most advanced weapons platforms in the world. Yet in the 1970s, officials began to note inexplicable problems. These included chunks missing from electrical cables, damaged sonar domes, leaking oil lines, and kaput sound probes. It turned out that sharks were attacking the subs. It was not that the sharks were inherently antagonistic to the military industrial complex. They were just used to going after passing whales, about the size and shape of a submarine, and the neoprene housing over structures like radar domes had a similar density to whale blubber. There are records of bloody divots being corkscrewed out of human legs too, although so far, no bites have been fatal to humans.

In 2023, cookiecutter sharks sank an inflatable catamaran in the Coral Sea, presumably due to another case of mistaken identity. That was the official story, anyway. Is the natural world trying to send us a warning? "Be good—or we will come and get even!"

Seeing It for Yourself

Will we ever have ocean tourism that centers on the night half of the aquatic experience? Dive boats do this already with organized night dives, and some advanced marine biology classes have nighttime field trips. The first octopus I ever saw in the wild was spotted at night in Baja during a biology group trip.

This bigfin reef squid is also called a glitter squid. They come closer to shore at night.

Separate from those events, what about experiences offered to the non-specialist, similar to how you can rent a bike for the day at the beach or go skydiving strapped to an instructor? This kind of safe, managed event seems overdue. Many things now are commonplace that seemed impossible previously. In summer there are now professional kayak safaris in the Los Angeles River, which sounds grim but is green and splashy and surprisingly fun. You can paddleboard across Walden Pond just outside of Thoreau and Emerson's Concord, Massachusetts, home, or go llama trekking in Rocky Mountain National Park, Colorado. I've run class-five rapids on the Nile in Uganda—watch out for the crocodiles!—while another friend, just shy of eighty, has sailed a hang glider across Rio de Janeiro. She told me she started on a mountaintop, landed on a beach, and in between she waved at office workers in downtown skyscrapers as she circled past, riding thermals like a condor.

So, with all of this increased sense of fun and outdoor access, it just seems that somehow, somewhere, somebody should be figuring out how we can better enjoy the ocean after dark. I don't want to do it as a scuba diver—no certificates and rebreathers for me—but could it be accomplished by using a small submersible? A glass-bottomed boat? Some kind of safe and guided snorkeling drift-swim? Have a look at this exquisitely glittery squid, please, and tell me that somebody is planning an adventure holiday that will include this species among the other natural attractions.

Tomorrow's Oceans

Traditionally, a chapter like this would be expected to end in bad news. Instead, let's just agree to set aside coral bleaching, overfishing, acidification, ghost nets, El Niño, microplastics, and pirate whalers. We will not allude to the vaquita, a dainty and utterly doomed Sea of Cortez porpoise down to the final ten individuals.

A pink sponge grows on mangrove roots, as land and sea meet in a productive contact zone.

Oil spill: two words that will not get studied here, nor the longer, more frightening version: "Russian tanker oil spill in the Norwegian arctic," which, when it inevitably happens, will manage to combine an ecological disaster with a political one.

There is so much bad news that neither one book nor one lifetime can accommodate it all.

One way through the list of depressing topics would be to pick just one small thing, like arranging a monthly beach clean-up or sponsoring a whale-watching trip for disadvantaged kids, and focusing your attention (and your money) on that. One person *can* make a difference, and so why not be that one person on your block, in your school, at your mosque or church or temple? There is plenty of bad news to go around, but I will say that, so far, the ocean ain't licked yet. We've done a lot of bad, stupid things, and yet here we all are: The oceans still function (surprisingly well) and the birds, fish, and squids are all (mostly) still there. Right on.

Global warming is probably a bigger problem than microplastic, and microplastic is probably a bigger issue than coral bleaching, but since none of us can fix those things easily or soon, for now all you can do is act ethically and model good behavior, and try to keep your small piece of the world running smoothly and well. For me, I think we respect life by celebrating it and honor nature best by approaching it with humility and curiosity. I still have not seen all the species of squid that I am interested in, let alone taken pictures of them or done sketches in watercolor. I find encouragement in the size of the task still pending, and I want to say, "Come join me—let's go see it together."

Fireflies, Foxfire, and Phosphorescent Waves: Things That Glow in the Night

One thing that has always surprised me is how many kinds of glow-in-the-dark Frisbees there are. As a child, they just seemed flat-out cool to me, and I must have owned four or five. Now flying discs even come with LED in-flight lighting, so here's a small toast to the miraculous improvement of everyday life. Of course, many parts of nature glow at night, too—from mushrooms to the dancing sparks of fireflies to the waves of the ocean itself. In this chapter we will explore this topic in all its quirky diversity, which will include talking about the two auroras, bioluminescent plankton (also called phosphorescent waves), fireflies and glow worms, foxfire (bioluminescent fungi), ice glow and earthshine, lichen, moonbows, red sprites, rocket launches, swamp gas (also called *ignis fatuus*), and that elusive sky show, zodiacal light.

Galileo named the aurora by combining the name for the Roman goddess of the dawn, Aurora, with the Greek name for the north wind, Boreas. This aurora fills the sky above Denali National Park, Alaska.

Some of these you may have already seen yourself, some you may hope to see some day, and maybe one or two you won't have heard of before. That's all right. The world is greater (and stranger) than we can ever imagine. Or to quote Hunter S. Thompson, "The world is still a weird place, despite my efforts to make clear and perfect sense of it."

The topics here are arranged alphabetically, so if you want to skip around, feel free to pick your own path.

Auroras

On earth we have two manifestations of the same phenomenon. The aurora borealis occurs only in the northern hemisphere and only at night—or at least, it's only *visible* at night. Also called the northern lights, this show consists of ribbons or curtains of ethereal light, usually green or yellow but also pink or red. The light seems to dance or shimmy across the night sky. The light might grow in intensity for a few moments, then fade, then slowly come back; it might hang still; or it might swirl and dance, the energy dial turned all the way up. A strong auroral event can light up the landscape brighter than moonlight.

Aurora australis is the southern hemisphere version—also called the southern lights. The plural label of both together technically is "aurorae," but most people just call them "auroras."

The auroras happen because the sun continuously streams out gases made of charged particles. These particles, also called ions, create the solar wind. As the solar wind approaches Earth, it meets our magnetic field, which has poles near but not at the geographic North and South Poles. Without this magnetic field to protect us, the solar wind would blow away our atmosphere, which would be "lights out" for life as we know it.

Most of the solar wind is blocked by the Earth's magnetosphere, and the sun's ions, forced to go around the planet like water parted by a boulder in the middle of a rushing stream, carry on deeper into the solar system. Not all of them flow past us, however. Some of the ions get folded back over themselves. Depending on the force of the solar wind, this can result in glowing blue or green halos around the Earth's north and south magnetic poles, causing an aurora. This is all happening sixty or even several hundred miles above the surface.

The display's colors vary because the atmosphere varies. While most of the atmosphere is made up of nitrogen and oxygen, there are also trace amounts of argon, carbon dioxide, neon, helium, ozone, and water. The densities of these gases change with altitude, as do the ways in which they interact with the solar wind. Red aurora colors come from interactions that occur higher up than the ion-atmosphere interactions that produce the greens, though to viewers on the ground, both colors look like they are filling the same space, "the sky."

The sun has an eleven-year cycle of increasing activity that gradually peaks and then lessens; years with lots of strong, frequent solar flares generate more—and more vivid—auroras than the quiescent years. During years with solar maxima, there are more sunspots, too, and there can be especially vigorous geomagnetic storms. Some mega-solar tsunamis, such as one that happened in late August 1859, provide globe-spanning auroras. That year telegraph lines failed, and for a few days the aurora was visible in such "non-polar" locations as Cuba, Hawaii, Mexico, and northern Australia. Our next predicted solar peak will be in 2025; how dramatic the auroras will be then is impossible to guess, but if you wanted to plan a vacation to Norway, Iceland, or Alaska, your odds would be better than average for a spectacular northern lights event that year and in the year just after.

The southern lights are harder to investigate. The best chance for seeing good aurora australis events would be to be one of the fifty brave souls overwintering at the US base at the

South Pole; the second-best place to be is aboard the International Space Station.

A vivid aurora (in this case, an aurora australis) dances beneath the windows of the International Space Station.

You can see auroras during moonlit nights, but most people find that a dark sky, away from the light pollution and distraction of cities, provides the best experience (and the best photographs). Long exposure times can make a photographed memory even more vivid than the experience was in person. Be sure to pack a tripod, handwarmers, and backup batteries. Cold temperatures drain batteries much faster than what you're probably used to at home.

Bioluminescent Plankton (Phosphorescent Waves)

In the movie *Apollo 13*, Tom Hanks, playing astronaut Jim Lovell, tells the story of how his navigation system failed during a night mission over the Sea of Japan. He had no lights and no homing signal. He ended up finding his way back to his aircraft carrier by following its wake, what in the movie he calls "a green trail, leading me home." He was describing bioluminescent plankton, which can glow bright green in the wake of a passing ship.

It works like this: Marine creatures, including fish, squid, tiny crustaceans, and algae, produce bioluminescence to either confuse predators, lure prey, or attract potential mates. As the density of a gathering of phytoplankton increases—usually in warm, calm conditions—the lights "turn on" if triggered by a disturbance, such as from a breaking wave or a passing ship. During the day, this density of phytoplankton looks red; that is the part called a "red tide" (even though it is not tide-dependent). At night, this concentration of organisms glows. Most often, this is due to a bloom of unicellular *Noctiluca scintillans* algae, which release light in response to external stimuli, such as the coastal waves throwing them from side to side. Currents, surf break, ships, dolphins, and even ordinary fish can also trigger the glow. You can surf in phosphorescent waves, which adds a magical, glow-stick-at-a-rave element to one's water experience.

To create the luminescence, the same chemical used by fireflies is involved. Responding to a change in pressure (the "disturbance"), there is a drastic reduction in pH within the cell caused by an influx of protons (positively charged subatomic particles) from the oxidation of a molecule called luciferin. This chemical shift happens incredibly fast—within twenty milliseconds—and creates a short burst of visible light that lasts for a tenth of a second. Yet that is it; it has spent all its fuel. It will be dark the rest of the night. For the waves to glow, others around it have to keep reacting. That the glows last so long into the evening shows us how many millions and millions of the algal cells are in a given bucket of seawater.

A related but rarer event is when the entire ocean is lit up without disturbance, in what is called a milky sea. This is dimmer and whiter than the neon of the breaking waves, and it can extend for miles around a ship at sea. It is observed most often in the northwestern Indian Ocean. Herman Melville, in his 1851

epic *Moby-Dick*, talked about a sailor's "silent, superstitious dread" on entering a "midnight sea of milky whiteness," as if "shoals of combed white bears were swimming round him." A description of milky sea also appears in the Jules Verne novel *Twenty Thousand Leagues Under the Sea*. A modern sailor, Sam Scott, says that when he experienced a milky sea, he already was very familiar with the usual blue-wave phosphorescence. "But this was not that. This was the whole of the ocean, as far as I could see, glowing a uniform, opaque green. Despite the compass still wheeling in its mount, the light in the water created an optical illusion, making the sea appear perfectly calm, as if we were gliding through phosphorescent skies rather than roiling seas." He sailed that way for four hours, crossed a boundary current, and was back in regular water.

Satellite imagery lets us know that milky seas can last for weeks. The science seems to be this: (a) bioluminescent algae colonies on the water's surface bloom and die; (b) as the dead algal cells rupture, they release lipids subsequently consumed by bacteria; (c) these bacteria multiply and eventually become concentrated enough to produce a continuous glow. How dense this is vertically—that is, how deep into the water column it extends—is unknown. Sam Scott continues: "Considering scientists believe it takes upward of one hundred million bacteria per cubic centimeter of water to begin glowing, the answer to this question [of how deep the phenomenon extends] could change the estimated number of bacteria involved in a milky sea by billions of trillions, or possibly even *trillions* of trillions." According to *Scientific American*, "One 2019 event, detected just south of Java, was visible for at least forty-five nights and covered almost forty thousand square miles—an area the size of Kentucky."

Fireflies and Glow Worms

A firefly is a beetle whose abdomen shines bright yellow. Also called lightning bugs, fireflies comprise two thousand different species, generally found in warmer areas rather than colder ones and in marshier places than dryer. Most are about an inch long. They go through the same larval stages as any other beetle, transitioning from eggs to in-betweens to their final adult forms. Some adults eat pollen, some are predatory, and some don't eat at all. Not all of these species glow in the dark, though what one then calls a lightning bug that doesn't have any lightning seems like the setup for a bad joke. Even if adults don't glow, all fireflies glow as larvae, albeit more dimly and via a different organ. Light-emitting probably started out the same way that bright stripes developed on toxic caterpillars, as some version of being able to visually say, "Stay back: I taste bad!" In the adults, it has transitioned from that into being a mating signal.

Fireflies are sometimes called glowworms, but that's not quite right. Most insect lovers distinguish between fireflies, which glow as adults to attract mates, and glowworms, which are larval-like females or actual larvae. Most glowworms, as with all fireflies, are related to beetles; none are true worms. The term

glowworm includes members from many different families of insects, including the alarmingly named fungus gnat, best seen in caves and damp places in New Zealand.

A kind of "See the Glowworms Here" tourism industry has grown up around the Waitomo Caves in New Zealand, where these glowing gnats are most easily observed. Besides their blue-green glow, these cave specialists are odd insects indeed, at least by our usual North American expectations. The gnat larva spins a nest out of silk on the ceiling of the cave and then dangles down threads covered with small, sticky droplets. Like a spider, they are trying to catch things to eat. The glow presumably helps attract food to their alluring strands. Their prey includes midges, spiders, and other invertebrates. If the glowworm feels a "hit," it reels in the line by ingesting the snare, and once the prey reaches its mouth, the gnat eats it alive.

With all fireflies, the light-emitting part is the lower abdomen. How the light gets made involves chemistry that, on paper, sounds complicated but isn't. To produce light, the enzyme luciferase acts on luciferin, in the presence of magnesium ions, adenosine triphosphate, and oxygen. Oxygen is supplied via an abdominal trachea or breathing tube. We now have light! From there the on-off pattern varies by species, and it can be anything from a full Morse code set of complicated flashes all the way down to your basic "always on" mode.

Sex can be risky, even for insects. Female *Photuris* fireflies mimic the photic signaling patterns of the smaller *Photinus*, which attracts males of that genus, which the larger females then eat. Besides the visual signals of the yellow flashes, fireflies also release pheromones to help attract and retain a mate; non-light-emitting fireflies use this method exclusively.

Humans have an infinite capacity for creating meaning out of the world, and the messages fireflies have been purported to symbolize outnumber the species of extant fireflies themselves. Among other things, fireflies supposedly represent the transitory nature of life, the souls of dead heroes, the arrival of planting season, the role of wisdom on a spiritual journey, the life and sacrifice of Jesus Christ, the value of creativity, or the brevity of summer. Feel free to insert your own interpretations into this list. Many songs reference fireflies, and Joss Whedon created a science-fiction series based on the metaphor. His version was sort of a Western sci-fi series, with horses, sidearms, and "firefly-class" spaceships; it culminated in a feature film.

Fireflies at Ochano-mizu, 1880, is a charming woodblock print by Kiyochika Kobayashi. He captures the serenity and magic of a warm night filled with fireflies.

Foxfire (Also Called Fairy Fire)

Bioluminescence in fungi is called foxfire—no relation to the search engine, Firefox. Of the one hundred thousand species of fungi, about eighty glow in the dark. Intensity varies, but for example, *Omphalotus nidiformis* (or ghost mushroom), the one shown on the next page, can be extremely bright. This species looks white in the daytime and grows on decaying wood, but it also can survive by accessing the general decayed matter in soil, in what is called being a saprotrophic feeder. It is poisonous, so does glowing at night help warn would-be consumers not to get involved? Another theory for bioluminescent fungi more broadly is that they are trying to attract insects to help spread their spores. Or it may be both, or both plus an unknown third thing—no single answer precludes the others from being true at the same time.

The "eighty" cited previously is probably an undercount. In New Zealand, for example, they are only beginning to search forests systematically, and multiple species of fungi are now known to be luminescent. Anna Chinn describes going out without even a headlamp switched on (so her eyes would be the most fully night-adapted) and discovering something startling, but in a good way. "One of the highlights of my life was noticing what I will call an actual ghost on the forest floor. I had shuffled into an area that appeared to be all blackness and was gazing down at the approximate vicinity of my feet. I detected what could have been a dapple of moonlight on a leaf and bent to pick it up. Since the moonlight moved with my hand, I knew it was not a dapple at all but a glowing, decaying leaf." What was going on? Chinn explains, "A fungus had colonized and consumed this leaf, and in doing so had taken on the leaf's form, and was now emitting light with the energy it had got from eating the leaf. If a ghost is the remnant energy and/or outline of a thing that is dead, then this is precisely what I was holding in my hand. A leaf ghost. Now I too was seeing spirits in the night forests of New Zealand."

▲ The well-
named ghost
mushroom
glows in Olney
State Forest,
Australia.

◀ The basket
fungus is
endemic to
New Zealand.

This story resonates with me, since in 1985, during a sixty-day trip backpacking across New Zealand, I came across a deliciously strange and endemic mushroom, the basket fungus. It looks like a small soccer ball that at the same time wants to be a TIE fighter from *Star Wars*. Maybe it looks like somebody's dome house that is under construction, since it is just the support struts, with no mushroom flesh in between. I later learned that one Maori name for it is "ghost poop."

In the days before smartphones and instant access to the all-seeing, all-knowing internet, one's field identifications relied on good notes and a good memory. I couldn't remember what this specimen was called, but I knew I had seen it in a library book once, in some kind of "mushrooms of the world" collection. The strange thing is that in my memory, it glowed in the dark. Now, it just could have been dusk, and the mushroom could have been glowing just in the sense that a white object shows up well in a dark forest. Or maybe I was using a flashlight to find the trail and spotted it that way. It does not appear on lists of bioluminescent fungus species, so maybe my memory is just plain wrong.

The one certain thing is that the more we look, the more we see, and that for the forests of New Zealand (and all the other parts of the planet as well), we have much yet to discover about the fungal world.

Ice Glow and Earthshine

Of the two hundred–plus moons and dozen dwarf planets in the solar system, four objects will be in the news often in the coming years. They are Enceladus (a moon of Saturn), Europa (a moon of Jupiter), Ganymede (also a moon of Jupiter), and Ceres (a dwarf planet located between Mars and Jupiter). The first two definitely have, and the second two probably have, large liquid oceans protected under thick crusts of surface ice. Other solar system bodies have water, too, including our own moon, but in all the other cases, the water is dense ice hidden from the sun in places like the darkest, most shadowed craters at the poles.

An illustration created by NASA shows how the nightside of the moon Europa glows blue from the radiation given off by nearby Jupiter. Beneath the glowing ice is a salty ocean of liquid water.

Enceladus, in contrast, has a huge ocean that is kept liquid due to the proximity of Saturn. As Enceladus circles Saturn and interacts with other large, nearby moons, it is stretched and pulled. That results in something called tidal flexing. This flexing produces a lot of friction and heat, enough heat to keep water liquid beneath the top layer of ice. Some water occasionally geysers to the surface, which allows passing spacecraft to measure what the water is made of. And the short answer is that if water equals life, and if life requires a certain default mix of chemicals, then the water on Enceladus and Europa offers places to look for life outside of Earth. We know that necessary ingredients for life are present—not just water, but also carbon dioxide, ammonia, salt, and other compounds. Meanwhile, as comparison ecologies, we know that at sulfur-rich vents deep under our own seas, life thrives in seemingly hostile (even impossible) conditions. You do not need access to sunlight to have ocean ecology; deep-sea hydrothermal vents host complex communities fueled by dissolved chemicals. Chemosynthetic bacteria and single-celled archaea found near hydrothermal vents form the base of an abyssal food chain, supporting organisms such as giant tube worms, clams, limpets, and shrimp. We also know that life survives even under the ice in Antarctica, where because of the salt levels, water stays liquid below the normal temperature of freezing. Even in twenty-eight-degree water, fish swim and starfish thrive—up to thirty-five species of starfish, in fact.

These facts swing us back to a central question. On the moon Enceladus, given that there is water there, and given what we know about extreme ecologies on Earth, could there be life in the salty slush under all the ice? In our lifetimes, we may find out the answers.

Until then, these ice moons are putting on a pretty and informative show. A recent NASA article explains how: "As the icy, ocean-filled moon Europa orbits Jupiter, it withstands a relentless pummeling of radiation. Jupiter zaps Europa's surface night and day with electrons and other particles, bathing it in high-energy radiation. But as these particles pound the moon's surface, they may also be doing something otherworldly: making Europa glow in the dark."

This light show has scientific value. "New research ... details for the first time what the glow would look like, and what it could reveal about the composition of ice on Europa's surface." NASA continues: "Different salty compounds react differently to the radiation and emit their own unique glimmer. To the naked eye, this glow would look sometimes slightly green, sometimes slightly blue or white and with varying degrees of brightness, depending on what material it is." To calibrated instrumentation, the glow would be not just attractive but deeply informative, letting us understand better how the layers interact and what the ice is made of. Forthcoming missions will investigate Europa and help us find out which worlds, besides ours, are aslosh with glowing saltwater.

Lunar glows happen a second way, too. *Earthshine* is the term for the way our own planet reflects light back onto the nightside of the moon. During a full moon, the moon is lit by sunlight; if you're standing on Earth and facing a full moon, the sun is behind you. During a new moon or when the moon is just a slim crescent, the sun is behind the moon or a bit to the side; the moon is backlit and should be a solid, dark disk. Yet we can still make out the faint, textured glow of the surface of the moon, especially using binoculars or a birding telescope. There is the crescent part, very bright, but the "face" has color and presence, too. That is because of the sunlight being reflected off our own planet onto the moon's surface. According to one farmers' almanac site, "this was called 'the new moon in the old moon's arms.'" I've never heard that said myself, but I like the poetry of it.

If you have access to a birding scope, the crescent moon is a good time for a bit of casual moon-watching. Full moons are so bright they blow out our night vision. A crescent moon doesn't do that, and as you scan the boundary line between light and dark, features like crater edges stand out in sharp relief. You can see the topography so well that some of the edges look sharp enough to cut a finger on. I raised my kids by making sure they all had a chance to use the telescope often, and when she was a toddler, my daughter called the sliver of a new moon a "sharp moon," which seems to me an accurate description.

As our cameras and our imaginations expand off-world into new environments, our language and our perception may need

Pearlescent clouds of frozen carbon dioxide shine brightly in the thin Martian atmosphere.

to evolve. On Earth we have late-night clouds, way high up, lit by sunlight long after the land below has gone dark. They look pearly, almost unearthly, and are called noctilucent clouds. When we finally go in person to Mars, those colonists will be able to experience clouds created by a similar process. The difference is that on Earth, our clouds are made of water, while the cold temperatures and low pressures of Mars create clouds that are a mix of water, ice, and carbon dioxide. They will shine differently than anything we're used to. Our vocabulary will need to expand, and landscape painters on Mars will have new effects to try to portray.

Lichen

If you have access to a UV flashlight (like the one you may have been using to look for scorpions), keep it handy for a second purpose. Some (but not all) lichens fluoresce under a UV light. Lichen is that crusty kind of pseudo-plant that grows on trees and rocks; officially, it is a symbiotic union of fungus and algae. Lichens are not bioluminescent like other things in this unit, but under UV light, some glow yellow, white, pink, green, or blue. Not all do, but enough that it's worth trying the next time you are out, and you can see it everywhere from Canada to Berkeley to Costa Rica. According to *Fungi* magazine, "More than a thousand substances have been detected in lichens, and many of them have properties that make them interesting for applications in pharmacy and medicine or to produce perfumes and dyes. What they actually do in the lichen is not known in many cases, but certain compounds act as sunscreens, protecting the lichen alga or cyanobacterium within the lichen from too much UV radiation that could damage the photosynthetic apparatus."

Looking for glowing lichen is what caused scientists to learn that there are glowing squirrels. In North America, there are

three species of flying squirrels, and each of the three is nocturnal. They, too, glow under UV light, as was discovered when a
flying squirrel sailed past a lichen team using UV lights and it was
glowing hot pink. A check of museum specimens showed that all
three do this. As mentioned earlier in this book, the platypus of
Australia glows under ultraviolet light; its fur shines with a mix
of green, blue, and purple. Theories to explain this suggest mate
selection, somehow staying opaque to the crustaceans that they
hunt, or the idea that perhaps the glow is a remnant of a previous
adaptation that lingers on, no longer active and yet still buried in
the genes. Expect to hear more about this in coming years.

Moonbow

Rainbows occur when sunlight enters water droplets and bends
(or "refracts"); they can only happen during the day. Moonbows
are the same thing, except they happen at night. Since moonlight is less intense than daylight, and since a full moon only
happens once a month, moonbows are fainter and less common
than rainbows, though part of that "uncommon" status may be
because fewer people are out looking. To an in-the-moment
observer, most moonbows look white or blue-white, although
taking a long-exposure photograph of the scene can help make
the moonbow look more saturated. (This is true of photographing the northern lights, too, as mentioned in that section.)

   We are so daylight-biased that the diurnal rainbows seem
like normal defaults, and moonbows seem like pale copycats.
It could have been the other way around. What if we praised
the moonbow as subtle and understated and derided daytime
rainbows as garish and vulgar? That has happened with how we
evaluate colors in other contexts. In the Middle Ages, when dyes
were rare and expensive, to have a scarlet cloak or a blue velvet
gown was a mark of wealth and taste. Painters saved their most

This shot captures two rare things at once. There is a moonbow but also an example of a Brocken specter, when the shadow of a backlit observer is cast onto a cloud or bank of mist.

saturated, indelible pigments for sacred subjects, like the robes of the Virgin Mary, which were often painted the densest shades of blue. Ultramarine, the most expensive version of this color, came from ground lapis lazuli from Afghanistan. Everyday peasants in the Middle Ages and the Renaissance wore browner and paler colors, since their dyes and primitive mordants didn't saturate clothes nearly as brightly.

Once bright colors became commonplace (thanks to modern dyes and mass production of textiles), there was a switch in social code, one that still holds true today. The loudest, brightest colors became associated with vulgarity, while modest and subtle shades like lichen green or eggshell white have become markers of class and distinction. You can see it in everything from fine linens to patio furniture: Understated color signals classy; bold color signals trashy.

Low class or high, rainbows and moonbows require sources of light to be aimed at just the right angle. The moon needs to be low on the horizon for a moonbow to happen, and it needs to be full or nearly full, and there needs to be water: a light veil of rain or the spray from a robust waterfall. Niagara Falls is a good site to try if you want to see it yourself, as would be Yosemite Valley in spring. The seasons matter in Yosemite; by late summer of low-snow years, the volume of water declines so much that at times the falls are not running at all. To know before you go, there are even online moonbow predictors, though I say just get in the car and take your chances, since one way or another something good will turn up. In winter, if Yosemite Falls is running well, you can stand late at night on the ice-covered bridge at the base of Lower Falls—be careful, it can be lethally slippery—and if the temp is low and the moon is high, you can stare up into a halo of swirling ice crystals. It is absurdly beautiful. Bears are hibernating then, but you might see a fox, deer, or ringtail. If you're with a companion, give them a hug and say, "Wow, what a planet. It sure is a good night to be alive!"

Red Sprites

A "red sprite" sounds like a cocktail you make with vodka, grenadine, and soda pop, but in meteorology a red sprite is a kind of fast, red, high-altitude lighting. The usual kind of lightning is white hot and goes from cloud to cloud, or from cloud to ground. In fact, a traditional lightning strike can be so hot that it fuses sand into glass, creating fulgurite, mineralized soil that looks like broken bits of reef coral.

Unlike regular lightning, red sprites are colder, higher, and start by shooting up, not down. Balls of ionized gas explode at great speed as high as fifty miles above the surface. To ground observers, they look like a burst of red flashes—although you usually need a high-speed camera or video to capture them. They last only a fraction of a second.

Red sprites were identified as early as 1886 but not photographed until 1989, just over a hundred years later. They remain poorly known now but seem more common during the peak and decaying stages of a massive thunderstorm, at a ratio (according to one photography handbook) of one per every two hundred regular lightning strikes. The types of storms most strongly associated with red sprites are "large mesoscale convection systems"; to see them well, you want to be fifty miles or even one hundred miles away and looking into a clear, dark evening sky. They seem to be too rare to need much further subdivision, yet types do exist. Their taxonomy is based on shape and color, so one study suggested that there are jellyfish sprites, columnar sprites, and carrot sprites. The last few years have generated more and more photographs of these events, and in some, the sprites do indeed look like Balrog-size jellyfish, each with a domed red "bell" and dangling, squiggly arms.

A storm chaser named Paul Smith lives in Canada and has captured more photographs of red sprites than anybody else. He even offers workshops, if taking pictures of the Milky Way or the auroras are options that are too tame for your photographic ambitions. If you went out with an expert, what would be the odds of seeing red sprites? He says that "some storms may produce only one sprite all night, whereas others may have a sprite a minute." This is an evolving field; as with many topics in this book, little

Red sprites spike over a supercell thunderstorm as lightning illuminates the cumulonimbus below.

is known and much is still left to find out. In documenting our observations, we all have a chance to contribute to the dialogue—or, to extend a metaphor from the oceans chapter, we have a chance to help fill up a few more gunnysacks of knowledge.

Rocket Launches

Some things look like nature but aren't. Is a jet's contrail part of nature? A contrail is just a long stream of frozen water droplets. The water is natural, but the specks of sulfur from the plane's exhaust that the water coalesces around are not, and without a plane passing high in the cold, thin air, the contrail (from "condensation trail") would never have happened. As people think about the divide between the natural world and the anthropogenic one, there is even a word for human-made clouds: *homogenitus*. The International Cloud Atlas gives this example of how to use that word: They suggest that a "cumulus cloud [that has] formed above industrial plants" would be known as *Cumulus mediocris homogenitus*. (Following Latin binomial tradition, this group capitalizes the "genus" of a cloud and lower-cases the "species" name.)

With rocket launches becoming more common, people in Florida, Texas, California, Kazakhstan, and the Gobi Desert (where China operates the Jiuquan Satellite Launch Center) will increasingly see strange, pearlescent clouds. Some of these clouds will bend at right angles or even look like a giant iridescent ribbon being tied into a giant bow. They are rocket contrails being lit up by the late evening sun, high up in the sky. The launches themselves may trigger the creation of real clouds, as the rockets insert water and pressure changes into the mesosphere, the part of the atmosphere where noctilucent clouds form. (Noctilucent clouds are high-altitude, very cold clouds that look ethereal and luminescent). Experiments over Alaska have

High-altitude winds above Vandenberg Space Force Base in California corkscrew a rocket's contrail into iridescent loops.

shown that the release of even small payloads of water can create noctilucent clouds within seconds.

This is nothing to fear from contrails like these, so far as we know now, but the sight is going to become more common, and it can be a puzzling view when you see your first one. It is a cloud but not a cloud; it is white but it is not white. (If the inside of an abalone shell could be turned into smoke, that is one way to think about it.) While most launches currently take place at only a few venues, such as Vandenberg Space Force Base in California, there have been successful launches from French Guiana, Iran, India, and the Marshall Islands, to name just some of the places where humans have reached for the stars. Additional new sites are planned.

After the 9/11 strikes, air travel in the United States and Canada halted for three days. One result was worry and inconvenience for tens of thousands of people. That was to be expected. A secondary, unanticipated result? Our skies were free of contrails for the first time since World War II.

Swamp Gas (aka Marshlights, Will-o'-the-Wisps, or *Ignis Fatuus*)

It is said that the devil takes many forms, and one thing he and his minions like to do is create lanterns full of false fire to lure late-night walkers deep into the fen, where they will be trapped forever.

There are many versions of this folktale, and the American tradition of carving jack-o'-lanterns at Halloween comes from it, too—the Celtic "Jack with the Lantern" character gave our candlelit pumpkins their odd name. The same legends also drive the scene in Tolkien's *Lord of the Rings* when Gollum, leading the hobbits Frodo and Sam through the Dead Marshes, sternly warns them not to follow the light, lest they end up trapped in the bog with the dead warriors.

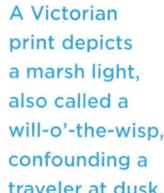

A Victorian print depicts a marsh light, also called a will-o'-the-wisp, confounding a traveler at dusk.

These stories combine our societal fear of the dark, the potential but real hazard of getting lost in a marsh where you can't see the horizon, and the fact that marshes do, at times, emit actual balls of flaming gas. My late father claimed to have seen one when he was growing up in North Carolina, and for him and his family, that feature was called swamp gas. Other sources refer to the phenomenon as marsh lights, *ignis fatuus*, or, most commonly, will-o'-the-wisp, in which "wisp" means a taper or twist of reed used as a candle or slim torch.

Marshlight comes from decaying vegetation. It is not linked to twilight in any physical way, but that is when it is just dark enough to notice it, and it's also when people would be passing marshes on their way home to the village. Saltmarsh wetlands, such as those around Norfolk in England, were places where reeds were harvested and bundled to make thatching for roofs and where salt hay was cut and piled up to be sileage for livestock. One hunted wildfowl in marshes or fished for eels. In the European tradition, a person might not want to live there full-time, but it was one of the ecological zones that was exploited for renewable resources.

As people worked in and around marshes, they encountered strange balls of light. That is because phosphine, diphosphate, and methane—all three of which are produced by organic decay—can cause photon emissions. And since phosphine and diphosphate mixtures spontaneously ignite on contact with the oxygen in air, only a small quantity of it would be needed to ignite the much more abundant methane and create an ephemeral flame. The equation then becomes fear + folk beliefs + methane = encounters with demons waving danger-fire. I think one more explanation factors in, too: Most rural people could not afford the high-quality eyeglasses that we now take for granted, and corrective LASIK surgery was of course not yet invented. As someone who needs both distance glasses and reading glasses, I wonder how much folklore arises from uncorrected vision.

No matter what the cause, for the preindustrial people who experienced these lights, the uncanny encounter was deeply disturbing. One detailed account comes from a Romantic-era poet, John Clare. A contemporary of Blake and Wordsworth, John Clare was a naturalist, a rural member of the working class, and a person afflicted at times with mental illness—though he was always an astute writer, in and out of the asylum. He is praised now for his direct observations of nature and for the honesty and forthrightness of his engaging voice.

John Clare's description of an encounter with marshlight follows, with some mild editing to update his quirky spelling and errant punctuation.

I determined to wait and see if it was a person with a lantern or a will-o'-the-wisp. It came on steadily as if on the pathway, and when it got near me within a pole's reach perhaps, as I thought, it made a sudden stop as if to listen to me. ... The luminous halo that spread from it was of a mysterious terrific hue. ... The darkness all around seemed to form a circular black wall, and I fancied that if I took a step forward I should fall into a bottomless gulf.

John Clare would have agreed with Gollum: Never follow the light.

Zodiacal Light

Earth rides on the ecliptic, which is to say, unlike Mercury and the planetoid Pluto, the Earth is on the same plane as the other planets in our combined circling around the sun. We all take a different amount of time to make the full circle, so while the Earth does it in a year, Neptune needs another 164 years past

The Milky Way spans the sky above the Cerro Tololo Inter-American Observatory complex in Chile. Zodiacal light is the pyramid-shaped wedge of white haze that starts at the bottom left and slants up into the center.

that. Even so, it's on our same orbital plane. Mercury is off from true by a few degrees, though still roughly aligned with the other planets; oddball Pluto is more off-axis yet.

This congruence to a shared planetary plane makes sense, given that the main objects in the solar system, including Earth and the sun, coalesced out of the same spinning disk of primitive matter. This was about 4.5 billion years ago, when the solar system was still a dense cloud of interstellar gas and dust. The dust cloud collapsed, possibly due to the shockwave of a nearby supernova. As it collapsed, it formed a solar nebula—a spinning, swirling disk of material. At the center, gravity pulled more and more material in. Eventually, the pressure in the core was so great that hydrogen atoms began to combine and form helium, releasing a tremendous amount of energy. With that, our sun was born, and it eventually amassed more than 99 percent of the available matter. The leftover bits became Earth, Neptune, Mars, and the other hangers-on.

In terms of its own vertical axis, the Earth is tilted to the side, true—that is why we have the seasons—but as far as our yearly circuit around the sun, we're aligned with the other members of the herd and we all share the same orbital plane, called the ecliptic. Something else that shares that plane is a disk of very fine interplanetary dust, the so-called zodiacal band or sometimes "cosmic dust," or more generally, zodiacal light. (The light is us seeing the dust when it is lit by the sun; the dust reflects light but does not generate any of its own.) This cosmic dust fills the inner solar system out to the inner fringes of the main asteroid belt, just past Mars. Each dust particle is only ten to three hundred microns across, so even the largest grains are less than half a millimeter. Most grains of beach sand are larger than that. It makes sense, then, that the dust band is going to be very faint, yet you *can* see it, and when you do see it, you're looking out at the actual space itself that fills the voids between the planets: This dust is distance made visible.

Zodiacal light is easiest to observe around the time of the equinoxes in March and September. Even then you can only see it from the darkest of dark sky sites when there is no moon; usually a few hours after sunset or before sunrise are best, depending upon which hemisphere you're in. Commercial astronomy apps and NASA's "What's Up" sky-watching blog and video series can alert you when and where to look.

Still, it *is* out there, inviting us to participate in the world in a new and different way. Brian May, lead guitarist with the band Queen, wrote his PhD thesis about zodiacal dust, finishing it 36 years after he had stopped working on it to be in a band. This proves (a) that it is never too late to go back to school, and (b) that even the smallest specks can generate profound thought and productive conversation. Once when Brian May was about to play a cover of the song "Dust in the Wind," he talked about his perspective with the audience. "We are made from dust—literally," he explained. "And we are made from the dust which comes from stars, from supernovas. As stars die, they give birth to the dust which gives birth to us, every one of us. That is what we are—dust in the wind."

The Starlight Smells Like Music: A Rainforest Case Study

L et's go for a walk.

We have considered undersea ecology and high-altitude storm clouds in the last half of the book, plus some odd bats and elusive felids, but let's set those aside now and go for a walk. What is it *like* to be out at night in the rainforest? We can explore that together now.

We start in Soberanía National Park, Panama, which preserves eighty-six square miles of lowland rainforest between Panama City and the Panama Canal. Because of birding hotspots like Pipeline Road, this park is a top destination for bird-tour companies. During a competition called a Big Day, 357 bird species were seen and heard here in twenty-four hours. Specialized snake and frog tours come to the park also, and insect lovers do very well, too.

If their itinerary includes a canal transit, some visitors arrive by cruise ship. For those who are land-based, places to stay include a five-star resort; more modest rentals in the town of Gamboa; and the

A flash lights up a violet-bellied hummingbird inside the forest.

286 ● The Starlight Smells Like Music

famous Canopy Tower, a former radar installation at the top of Semaphore Hill that has been converted into a treetop eco-lodge. All of these choices are equally good for seeing sloths, howler monkeys, parrots, and blue morpho butterflies, and all choices (especially Canopy Tower) are also good bases from which to carry out night excursions.

Night walks do require a bit of setup ahead of time. At the front desk, I usually want to know two things: When does the kitchen close for the night? And secondly, if I go out, will there be locked gates later, and if so, how can I get back in? Some lodges want you to take a local guide, but to be honest, that can produce mixed results. Some of the most brilliant daytime guides are of less use at night, since they may not have much experience and could even have the same fear of the dark that many "civilian" people have. I've even been told by a guide that there is nothing around at night (when I knew that not to be true), and once, standing on a bridge late in the afternoon, I was assured there were no bats nearby, when just ten minutes before I had seen them roosting on the underside of the same structure we were standing on.

For any night walk, you're looking for field companions with good woodcraft; there is an art to walking quietly and managing not to talk, talk, talk all evening, and you also want people who can be alert to subtle signs. You might find a frog or snake based on a small flicker of movement, so paying attention does matter. An experienced traveler might bring a laser pointer, in order to create a green or red dot to help others locate the correct tangle of vines that holds the hidden opossum. It helps if everyone in the group has a flashlight but can agree ahead of time which person will control the main beam and who among the rest will operate with their headlamps on the red setting. (Red lights bother animals less, and a red headlamp attracts fewer moths.) In comparison, ten loud people with ten bright flashlights won't see any animals at all.

Our walk will take place during the boundary month between the wet season and the dry. For me, my "night" starts in the afternoon, not only to arrange for meals and top off my canteens, but also to enjoy the transition as the day ends and the evening eases into existence. On this particular day, there had been a thunderstorm earlier. As the front hit, it set off both car alarms and the hooting of howler monkeys. Late-in-the-day rain in the tropics can cool things down and stimulate a late-afternoon bout of feeding and vocalizing among birds and other wildlife.

Some of the area hotels have good views over the tops of the forest trees, so you can get views that are hard to have when down on the ground, trying to crane your neck to look straight up. Standing on Canopy Tower's upper deck, I scan the sky and horizon with my binoculars, picking out swifts, Geoffroy's tamarins, tanagers, and passing parrots. I can hear a tinamou, a ground-dwelling, pigeon-turkey kind of bird whose mournful, quavering trills carry a long way through the forest. Ah, here we go: Here is something good, an eye-level, late-afternoon sloth. Two species are possible. This one must be a three-toed sloth, with its yellow face and black commas across the eyes. The two-toed sloth is larger and expected deeper in the primary

A three-toed sloth feeds at sunset in a barrigon tree.

forest. That one is also more closely related to the extinct giant ground sloths than the three-toed kind are. Both sloths have long, sharp claws—climbing hooks and weapons, all in one unit.

My first attempt at a photo of the sloth just produces a black silhouette, since the camera is tricked by the ambient light. It's a nice image of a pastel sunset but not a nice picture of a sloth. Attempt number two with better settings and a switched-on flash helps bring out the animal's details, as we see in the photo of the sloth above.

Sloths eat slowly, digest slowly, and poo slowly, or at least they poo infrequently, coming down once a week to defecate and urinate at the base of the tree they have been feeding in. Since algae grows in the microcracks of sloth hair, their coarse gray fur can look greenish, which adds to their camouflage. Moths and ticks live in the fur, too, so a sloth is an ecosystem unto itself, especially when you consider its unique gut biome that helps it digest foliage.

Something else to do before dark is to scout places that may host bats later in the evening. Many bats have a two-part feeding pattern. They are active early in the night, go to a midnight roost to rest "midday," and a few hours before dawn become active again. The night roost can be the same as the day roost or might be a place they only use while out foraging. In popular imagination bats are associated with caves, but as we have seen earlier, bats also roost under palm thatch and tree bark, inside banana leaves, or in and around human structures, including road culverts and abandoned houses. Once in Peru I was taken to see bats in an abandoned house that had a handmade spiral staircase linking the ground floor to the now unsafe upper level. The house was derelict—it had no furniture, no plumbing, and no glass left in the windows; half the roof was gone; goat poo and bat guano coated the plank floor; and the plaster walls were covered in pornographic graffiti. But then there was this oddly intact, utterly anomalous spiral staircase in the middle of the main room. It was hard to help catch the bats (needed as part of an acoustical study), since all I could think about was how good that staircase would look in my house in California, and how impossible it would be to ship it there.

Common big-eared bats cling to a vine inside a road culvert.

Following a tip, today I know where to look first. Common big-eared bats, like those at the bottom of the previous page, often hang in a cluster of four to six individuals cozied up in a bundle. According to Fiona Reid, "Food is usually consumed at a night roost. A collection of wings and legs dropped at a night roost in Costa Rica contained insects of thirteen different orders, with small scarab beetles, grasshoppers, cockroaches, crickets, and katydids predominating." In Mexico, butterfly wings were found at a roost, and in Panama, the body parts of flies and beetles. I mark this site so we can check back later, when there may be a new species mix present.

One final thing before we go out is to check the lodge's bug light. Many nature centers now have these, as a way of drawing in nocturnal insects with a UV light and encouraging them to cling to a draped white cloth so nature lovers can admire their color and variety. A tropical wonderland such as Costa Rica or Panama can have as many as twelve thousand different species of moths, and so for me, in terms of trying to guesstimate identifications, I often give up before I even start. I just say "moth," and walk away. This defeatist attitude does not speak well to my character, but my private ambition to know everything there is to know about everything hits a wall when it comes to neotropical insects. I like looking at them, though, and on the edge of the drop cloth I can see that a slender black-and-yellow bug with immense sweeping antennae must be one of the long-horned borers. If our bodies were the size of this insect's body, and if its antennae were antlers on us, we would each have a set of horns that sticks out eighteen feet on each side. Each of us would need a full-size tractor-trailer to get around, and even then, only if we stood in the middle of the trailer, facing the wall. Humans being humans, I am sure there would be some among us who would cut our antlers clean off, just to defy our parents, while there would be others who would pay extra for antler extensions.

Snapping out of those idle thoughts, I realize it's getting dark. I take a few photographs of a particularly leafy moth that is resting

next to a bullet-shaped, green-black beetle on the drop cloth, then decide it's time to gather my fellow "nocturnalists" and get going.

Walking down a road into the forest makes one aware of the instant drop in light level. Night comes earlier to the inside of the forest than it does outside the canopy. It is humid, more humid than I am used to, and it smells— well, it smells like a rainforest, which is to say, mulchy, rich, decaying, and alive. In the novel *Purity*, Jonathan Franzen describes the impressions of his main character Pip upon her arrival at her lodgings in the rainforest of Bolivia. Focusing on her perceptions in third-person prose, Franzen writes how she noticed that "two scents at once, distinct like layers of cooler and warmer water in a lake—some intensely flowering tropical tree's perfume, a complex lawn-smell from a pasture that goats were grazing—flooded through her open window. From a cluster of low buildings on the far side of the valley, by a small river, came a trace of sweet fruitwood smoke. The very air had a pleasing fundamental climate smell, something wholly not North American."

To me, the forest's smell is like the inside of an old greenhouse or garden shed, but with vanilla, rotting meat, and exotic flowers added in. It would not make a good perfume, but it does make a very good rainforest. Smelling it, I always feel a kind of homecoming.

Stopping to listen inside the darkened forest, I can hear the downed-wire *zzzzt zzzzt* of cicadas. At least, I *think* they sound like cicadas, although it could be another insect family entirely, and how would I know? The sound rises and falls in a hypnotic pattern. To biologist John Kirchner, this background din is reminiscent of the oscillating pitch of French ambulance sirens, *eeeee-ooooh, eeeee-ooooh*. There are a few mosquitoes floating past, though fewer than I have witnessed in city parks in the Midwest. I've already spritzed myself with eau de DEET, so when it comes to "mozzies" (as they're sometimes known in Australia), I am okay. Malaria and yellow fever—both spread by mosquitoes—used to be serious problems in Panama during the initial digging of the Canal, but they are nearly eradicated now. Mainly I just don't want the irritation of itchy bites in the morning.

One strange aspect of the closed-in forests is their microvortices of temperature and air flow. Maybe I am attuned to them more since I am hyperalert going into this unfamiliar habitat. Maybe it's because I'm already damp from the post-storm humidity, and my wet skin registers small changes with uncanny sensitivity. Whatever the cause, I can feel odd pockets of different air—not hugely different, but distinct even so. At times it feels like my lower body is wading through a pool of cooler air while my upper body inhabits warmer strata.

Oh, good—our first frog; or rather, looking closer, we see that it must be a toad. Large and warty with a stoic, slightly

Cane toads are native to the neotropics but have been introduced into Australia, New Guinea, and the Pacific. Their skin is covered with a poison that is toxic to dogs and other animals.

grumpy air, this one is a cane toad. It doesn't know it, but it is a much-hated beast in places like northern Australia. The cane toad was foolishly introduced there to control insects in the sugar-cane plantations. Now its population numbers in the millions. Native marsupials like quolls (page 164) try to eat them, but this toad is toxic, and some animals (including domestic dogs) are deeply susceptible to the secretions on the toads' skin. *National Geographic* calls this toad "one of the worst invasive species in the world." In this dog-free forest, though, they are native and just another cog in nature's machine. We take pictures and move on.

Leaving the toad to sit solemnly in his toad-ness, we carry on, scanning the trees with the new secret weapon of nighttime amblers: a thermal imaging scope. It detects heat, so either a resting sloth or an active termite nest will glow white-hot compared to the fuzzy gray of the rest of the image. You can make out other shapes in the display, dimly, and so can tell the road from the trees, but it's all a rather lo-fi visual experience—or at least it is for me, since I have one of the earlier models. The only comparison I can give is to refer to what Frodo's vision looks like when he puts on the Ring of Power, such as when the Nazgûl attack the hobbits on Weathertop. Frodo's world becomes surreal and out of focus, and colors and shapes no longer make precise sense.

Thermal imagers are expensive and evolving rapidly, so cheaper ones coming out now often have better resolution than models from even a few years ago. "Thermal imaging scope" is the correct, formal name, but most people call them heat scopes. They are slightly illicit, since some security screeners at airports are dubious about their purpose; to an inquisitive guard, an animal spotter's innocent monocular and an insurgent's night-vision scope would look like the same device. Looking up through one of these scopes is odd, since the sky, being without local heat sources, is just a featureless dark void—there are no stars at all, just blankness. That is very unsettling, at least to me.

However, with a tool like this, even an animal hidden behind foliage or back some meters into the forest stands out as a hot, white blob. That is how I have noticed a robin-size songbird roosting overhead. Using flashlights, we see that I have found a sleeping wood thrush.

This species nests across most of the eastern United States and winters in Central America. Is it a North American bird that winters in the tropics? Or is it a tropical bird that briefly leaves home, visits Missouri or Massachusetts, and scoots right back? In other words, from a North American perspective, is it "ours" or "theirs"? Either way, the species faces problems. Cornell University's *All About Birds* website says, "Though still numerous, its rapidly declining numbers may be due in part to cowbird nest parasitism at the edges of fragmenting habitat and to acid rain's depletion of its invertebrate prey."

Each part of the ecosystem connects to the next; the cowbirds' impact on thrushes starts because of a quirk of evolution. Originally, cowbirds were blackbirds that followed bison herds. Because they followed herds, they were never in one place very long, and so they laid eggs in other birds' nests and took off, leaving others to raise their young. This behavior is called brood parasitism, and if only done in one area briefly, as herds move through and then depart, it does not severely impact songbirds overall. In brood parasitism, the invading birds hatch first and outcompete native nestlings for food or even push them right out of the nest. Cowbirds also lay a *lot* of eggs, overwhelming the host nest in that way.

Now things have changed. The buffalo are gone and feedlots for cattle have taken their place. Feedlots stay put; they do not migrate across prairies. And, since cowbirds prefer open plains and we have cooperated by clearing forests, and since cowbirds can find grain and insects in cattle manure, they no longer forage briefly and move on, following herds as they depart. As static, year-round residents, they can impact local warblers and thrushes quite severely.

Birdwatching at night sometimes means encountering sleeping birds, like this migratory wood thrush.

Sweet dreams, little thrush. With luck, you cannot imagine the dangers you are yet to face on your return to the fabled lands of the north.

Once again, I take a photograph and we move on.

The storm earlier in the day has weakened old wood and saturated the heavy coating of moss and epiphytes that many trees carry, and so when a branch gives way and crashes down with a loud *wah-WHOOSH*, we all jump. It takes a moment for our brains to reassure our pounding hearts that a jaguar attack does not sound the same as a falling tree, and the odds of seeing a jaguar in the park are very close to nil. A margay or an ocelot maybe, but a jaguar? No. The incident does remind me that a sister branch to the one that fell could be directly overhead, and I should be attentive to any creaks that imply strained and rotting wood. The fact is, nothing is truly safe in the world, and while I am probably fine here in the night forest, I need to be wary. Snakes are always possible or there could be a car coming with no headlights, or I could even just turn an ankle in a pothole; safety relies not on others but on myself. As the old bumper stickers used to say, "Be Alert—the World Needs More Lerts."

Being alert pays off now, since *there*, what's that shine at the edge of the light? We use binoculars and can see that it's some kind of frog. We lean in closer. Oh, right on, it's Sylvia's frog, *Cruziohyla sylviae*. Who was Sylvia? Sylvia is not a "was" but an "is"; she is the granddaughter of Andrew Gray, a British herpetologist who in 2018 solved a one-hundred-year misidentification.

The short version is that a frog was identified in 1908 and everybody since thought they were seeing that frog, but the frog they were seeing was actually a different kind, not the 1908 frog. The frog these people were seeing was a new species. The 1908 one was still around but was different, and nobody noticed that there were two different species . . . at least, not until Andrew Gray figured it out. The 1908 frog lives in Ecuador. Gray's kind—the one we are seeing tonight—lives in Costa Rica, Honduras, Nicaragua, and Panama. It belongs to a group called leaf frogs.

"But wait," as the infomercial hosts exhort, "there's more." Win once, win twice, since we keep looking and nearby is an even better treat—rarer than the frog itself and visually so odd that it's hard to process what we are seeing.

We have found an egg mass for Sylvia's leaf frog, made up of vertical strings of gelatinous, pea-size clear eggs, and inside of each egg is a tiny, curled-up tadpole. Shining a light from behind reveals dozens of these baby-frogs-to-be, each in its fluid

After a century of misidentification, Sylvia's leaf frog became its own species in 2018. It is named after a scientist's granddaughter.

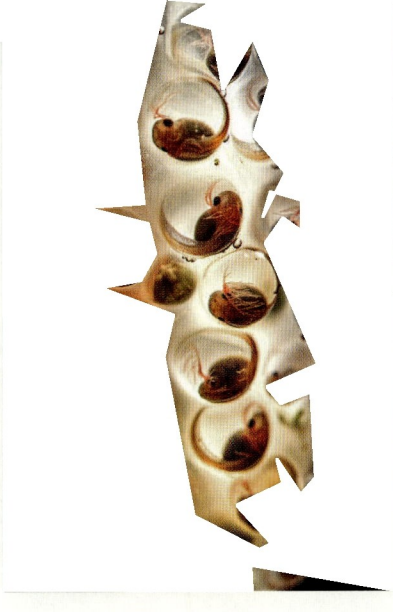

mini-womb. Being morbid and prone to fits of dystopia, to me the baby frogs recall the fetus fields in *The Matrix*, those endless rows of pods where reality-challenged baby humans are grown like crops.

To my friends, the frog eggs look like snow globes, pearls, or mesquite-tree seedpods, so I am glad that nobody else shares my nightmare visions. This speaks well of them and perhaps poorly of me. Nature: the ultimate Rorschach test.

Almost as a joke, somebody says, "Where there are frogs, there are snakes." It is true; some species of snakes hunt frogs as their main meal choice. We keep walking, scanning the leaves not in fear, but in hope. Snakes are cool, and several of us keep lists of all the snakes and other reptiles we have seen around the world. For me, recent additions include Komodo dragon and banded sea snake, both in Indonesia, and speckled rattlesnake in the eastern Mojave Desert. The heat scope picks up something interesting in a small palm. Not a snake—they are cold-blooded, so are basically the temperature of the air. The heat scope can't read them. And it can't be a peccary; they do not climb trees. Maybe a coati? A monkey? Heat scopes and flashlights reveal the truth: We have been lucky enough to come across a northern tamandua, a midsize anteater. For some reason, it doesn't react to us and keeps coming down its tree. After a quick flurry of pictures, we back off to give it room.

While the tamandua can be common, it is not that well-known. A different kind, the giant anteater, is the one most people are familiar with: bushy tail and extra-long snout (and mascot of my alma mater, UC Irvine—"zot!"). That kind is more typically seen in open grasslands rather than closed-in forests, though whether that is its own preference or just observer bias is hard to know. The tamandua species tonight is snouty, too,

like its big brother, but is midsize: it is a sensible Toyota Camry compared to the Cadillac Escalade of the giant anteater. In Brazil I have seen the big one and also this one's sister species, the southern tamandua. There is also a cute teacup sort of anteater, the smallest of all, called the silky anteater. If found curled up asleep in vines, it is often described as a furry, golden tennis ball.

The name "tamandua" is a Portuguese transcription of an Indigenous phrase that means "ant catcher" (though besides ants, they also like termites a *lot*).

Snakes are easy to see in the forest at night, so long as you don't look too closely the second time—so far tonight I have seen ten or twenty "stick snakes," but never the real thing. After another half hour we have some herpetological good luck and our snake-less night becomes officially snake-ified. And it is even a frog-chaser, just as predicted. We have found a slim, well-sighted, night-active serpent called the cat-eyed snake, named after the vertical slits in its large pupils. These snakes can be long—more than five feet—but this one is more like two and a half or three feet, sinuously riding a magic carpet of branches about four feet off the ground. Being thin and agile, it navigates the just-aboveground world of vines and bushes with weightless ease.

Humans have a natural fear of snakes, an instinctual caution that has been intensified by cultural descriptions such as calling an ethically challenged person a "snake in the grass," or saying of this or that lost item, "If it had been a snake, it would'a bit ya"— meaning it was so close at hand it could have attacked you. In this case, our evening find is a small snake and not large enough to harm me. It has weak venom and I don't plan to handle it. Since I am not a frog, a mouse, a lizard, or a cricket, I have nothing to worry about.

There are about four thousand species of snakes in the world. We now know that snakes evolved from lizards, though the fossil record is poor, and as with so many topics, more information would be gladly received. There's a birdwatching term that applies when you've seen a new species well enough to count it, but not well enough to enjoy it. That is the expression, "Better View Desired." A lot of the fossil record could earn that designation. In high school, one of my friends wrote a mock battle ode whose punchline was that a snake "is just a de-feeted lizard." As it turns out, he was more right than he knew.

Time for a rest stop. "Drink water—and then drink more." That's my safety advice for jungle travelers. Even at night you sweat a lot, or at least I sure do. My shirt is drenched. Resting with all the lights off we see much more than before, and among other things, looking up, we can appreciate the flickers of the moon and stars peeking down at us from gaps in the treetop leaves. The air is damp, rich, and cool. Somebody says, "The evening smells like music." I don't know what that means, but it feels true. More sounds lift into consciousness; an insistent creature about ten feet to my left is going *whit, whit, whit,* though if it is frog or insect or bird or ghost sharpening a dull knife, I cannot quite puzzle out. To cool off, I try fanning myself with my journal. That makes me drop my pen, which makes me switch on my light (turned to its dimmest setting, of course), which makes me realize that on the same wall where I am sitting, there is a lovely little moth, one of the many thousands and thousands I can't/won't/will never identify.

It is pretty, though, and I want a picture. Leaving the head-lamp on for a moment—I will need it to help the autofocus lock on the subject—I slowly ease my camera around on its cross-body sling, make sure the lens isn't flecked with mud, click on the flash unit's power, wait for the verification of its electronic *eeeeep,* and then enter in a reduced setting on the flash so that when I take a picture the light doesn't go off like an atom bomb.

Of course, by now the moth has flown off.

I decide to wait, more from laziness than patience—putting the camera away seems like more trouble than it is just to sit there, dazed and vaguely inattentive. A person with their camera out and switched on is clearly working, not loafing, and here's a tip for everybody—on long, uphill walks you can buy time for a longer break if you pretend to be taking landscape photos during the pause from hiking.

The moth comes back and I kneel down next to the wall, trying to keep its wings parallel to the camera's sensor plane, so more of the insect will be in focus. No satisfying "click," though—I am using one of the new mirrorless cameras, which I have set for silent running. There is no mechanical shutter to open and close; the all-electronic camera can take one picture or

An interesting moth lingers on a piece of wall, holding steady for the author's unsteady camera.

a hundred, all without making any noise. You can turn on an artificial shutter click in the menu, but I make enough noise as it is and certainly don't need more sounds that might scare wildlife away.

Chore done, camera put away, I try to catalog the sounds. A *creeeek creeeek* might be a frog. Other frogs are definitely chirping; cliché as it is, there is no better word. There is a faint, persistent mosquito whine—is it time to freshen up my repellent? We can hear dripping water everywhere, which reminds me that most tropical plants have "drip tips" at the ends of their leaves. This type of leaf ends in a pointed, downward tip, to help drain water. The air is starting to turn—will there be rain?—and tiny whorls of wind stir the leaves. I try to translate a firefly's dot-dot, dash-dash pattern into words spelled out in Morse code. I must be doing it wrong, since it seems to be saying, "Hey dummy, don't look now, but you're sitting in ants." Closing my eyes, I am impressed by the three-dimensionality of sound. We have insects in all directions, everything in full stereo and seemingly calling from under the ground, through all the strata of the forest, and up to the very tips of the tallest trees, one hundred feet up.

Now a pauraque starts up, very close, just down the road. This is a nightjar that goes *WOHEERRrr* (pause), then *wik-wik-wik woHEERrrrr*. I first heard it while camping in south Texas, on a very hot and sticky night forty years ago. Here in Panama, it's called *bujío*, though my phrasebook translates that as "booming," so there must be another meaning. Hearing it, and hearing that it seems to be staying in one place, I sigh and get up. Time for another picture. Earlier we had heard a species called rufous nightjar, which gives out a rapid, high-pitched *wik-wik-WHE-ooh*, like a North American whip-poor-will.

We never saw it, despite trying. I also think I can hear a common potoo, but it's so far away, I am not sure. Dr. James Bond, author of *Birds of the West Indies* and an actual person

whose name Ian Fleming appropriated for the spy books, transcribes the sad, plaintive call as a long, drawn-out version of "poor-me-all-alone."

James Bond the ornithologist knew his name had been repurposed for the novels and seemingly was fine with it. Fleming thought the name was Anglo-Saxon sounding and very masculine, even though the birdwatching namesake was American, not British. None of them could imagine the degree to which the name would become universal today. A hundred years from now, Agent 007 will be drinking martinis and chasing bad guys on the moon.

Besides water bottles, insect repellent, and crepe-soled friends, I recommend a fourth and final piece of an essential travel kit: a pocket notebook. I usually bring several sizes. One might be truly pocket-size and others go in a slot in my shoulder bag or backpack; those are usually 5 x 7 or 6 x 9. (At home I also use 9 x 12 sketchbooks as journals.) I know everybody reflexively reaches for phones these days, but I just will point out that phones break, passwords disappear, and hackers gotta hack. Number of journals I have had stolen? Nil. Number of times my journals have been wiped clean by an upgrade that went badly? Nil. Number of times I had to stop using my journal because I had no battery left? Again nil. I confess some of my pages are nearly illegible, but that's on me—I am often too lazy to put on my reading glasses and write by muscle memory and blurry guessing. As bad as it comes out sometimes, it never is as whacky as when I try to text without my glasses on. Between my finger-typing and autocorrect's absurd assumptions, I might as well be adding chapters to *Finnegans Wake*.

To fact-check this chapter, I later went back to look at my notes from the night in question. I do see now that I was in a bit of a pique that we had not seen the rufous nightjar, which apparently had been out of sight too high up for us to make out with our spotlights. We heard it but could not find it. Some birders keep "heard only" lists, but I am not one of them. You either see it or it doesn't count, and that is partly because some birds like mockingbirds are very good at sounding like other birds, such as the buff-collared nightjar in Arizona. And so, on a journal page of otherwise miscellaneous entries, I used a colored pencil to add (somewhat petulantly), "Hey, stupid bird, come out and fight fair." I also have quotes on that page from what I was reading that day and a sarcastic note about the weather, from earlier when it seemed like it would never stop raining.

Most field biologists create tidier, better-structured logs than mine ever are. Martin Moynihan—founding director of one of the most famous research stations in the world, the Smithsonian Tropical Research Institute—was studying bird behavior in the 1950s. Insight never goes out of fashion: His field notes still offer useful information today.

It is getting late; time to think about the walk back. Lightning on the horizon and rising wind mean that more weather is coming. Just when I think I am done with insects for the evening, a black witch moth (or a look-alike species) catches my attention. I also photograph something with large, green, swept-back wings that I later determine is a Pluto sphinx moth.

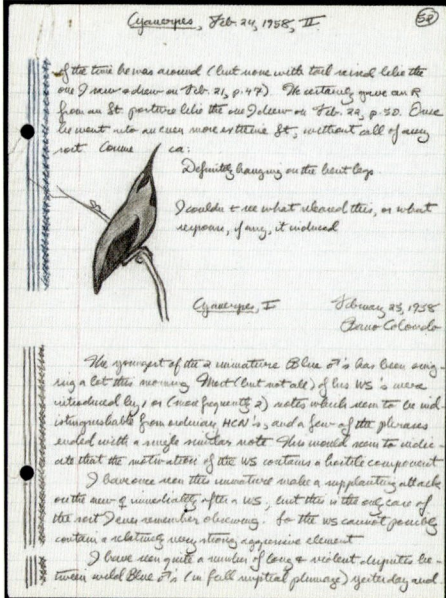

I forget about insects as soon as I hear "Frog!" This time it's a poison dart frog in clear view, so time for more pictures. This one (like most species in the dart frog grouping) is about an inch long. The ants and mites they eat make dart frogs toxic, and the frogs warn predators of this by having what is called aposematic coloration—a pattern of bright colors to warn things off. The "poison" part of the name is accurate, but the "dart" is less so; Indigenous people did and still do make poison from poison dart frogs, but only rarely and only by using a few out of the hundred-plus poisonous frog species. Most of the time, native hunters tipped arrows and blow darts in curare, an alkaloid derived from plants. Curare only interacts with the target animal's blood; you can safely eat a tapir (for example) that has been killed by a curare-tipped weapon, because the meat is not contaminated by the poison. In zoos, dart frogs lose their toxicity since they no longer are eating Panamanian forest ants; in captivity they are fed wingless fruit flies dusted with calcium powder. "Breakfast of champions," we can say without sarcasm.

▼ Field notes from Martin Moynihan, founding director of the Smithsonian Tropical Research Institute in Panama. This page is from 1958—few adults today write in such fluid cursive.

▲ A page from the author's notebook describes the rufous nightjar (among other things).

This green and black poison dart frog is smaller than a quarter yet has enough poison to kill a human. The ones kept in zoos lose their toxicity due to the change in their diet.

Soberanía Park attracts wildlife lovers from around the world, and our walk tonight has been interesting but not unique. Others who have come on prior trips have seen even more animals than we have. Russian ecologists Karina Karenina and Andrey Giljov have shared advice online from their own visit to Panama. They spent multiple nights in the park, using thermal scopes and handheld lights. In a report posted on the mammal-watching website mammalwatching.com, they have recommendations for teams who hope to follow their route.

They start with a basic point about all nature after dark. "Just walk slowly," they say, "and do not give up during prolonged periods of 'no mammals.' Some nights at Semaphore Hill were more productive than the others and the same is true for the different parts of a particular night. Once we made it the whole way down the hill from the gate at the top without any noticeable sightings (agoutis and howlers sleeping in the trees don't count for us), but the way back brought us three porcupines, two tamanduas, Tome's spiny rat, two woolly opossums, a paca, and a silky anteater!"

Their luck may not be your luck, and yours may be worse or it may in fact be better. The ocelot photo on page 183 was taken in Soberanía, on the birding route called Pipeline Road. Not many people see this exotic cat in the wild, but every year some people do—so why not you? If we ask ourselves what makes us human, one answer is that we get to celebrate what we love, not what we fear. It is a choice, a path, a possibility. We find beauty by allowing ourselves to create beauty—Terry Tempest Williams says this—and beauty turns up in odd places, even behind a rock or in the way Venus, the morning star, stands on its tiptoes balanced on the edge of a saguaro, as nectar-feeding bats and moth-nabbing poorwills fly around it.

The night is annoying, frightening, impossible. And yet—
And yet . . .

. . . it is also so full of beauty that I know I will never get enough of it.

Thank you for joining me on this journey.

I can't wait to see what we will all discover together.

Acknowledgments

For advice, text review, photography, and good company, the author would like to thank Charles Anderson, Autumn Anderson, Pamela Anderson, Karen Baker, Sacha Barbato, Gary Carter, Lisa Carter, Paul Carter, Carol Chambers, René Clark, John Eakin, Dodge Engleman, Bill Fox, Jonathan Franzen, Mike Guista, Jon Hall, John Haubrich, Abbey Hood, Amber Hood, Fred Hood, Marek Jackowski, Mathew Jaffe, Sebastian Kennerknecht, Michael Light, José Gabriel Martínez-Fonseca, Vivek Menon, Coleen Moloney, Christine Mugnolo, Zia Nisani, Bill Noble, Miguel Ordeñana, Carolyn Purnell, Avotra Rabearisoa, Fiona Reid, Mike Richardson, Peter Ryan, Phil Telfer, Ian Thompson, Erin Westeen, Sarah Winch, and Cal Yorke.

Photo Credits

All photographs by the author, except as noted.
We thank the contributors for their help.

Wikimedia

Alamy

Index

T

Charles Hood is a poet and naturalist. The author of twenty books, he has been a factory worker, a ski instructor, a dishwasher, and a nature guide in Africa. Nature study has taken him across all fifty of the US states and to eighty countries, from New Guinea to Borneo to the South Pole. Along the way he has been lost in a whiteout in the Himalayas, contracted (and survived) bubonic plague, and published more than seven hundred photographs. His titles with Timber Press include *Wild LA*, a field guide to reptiles and amphibians, and a guide to the best roadside hikes in California. Jane Goodall wrote the foreword to his book *Wild Sonoma*, and his essay collection, *A Salad Only the Devil Would Eat*, was named the Nonfiction Book of the Year by the editors of *Foreword* book review. He lives in the Mojave Desert with two kayaks, two mountain bikes, two dogs, and five thousand books.